Physical Geology Laboratory Exercises
(EEES 1020)

Third Edition

prepared by

James A. Harrell, Ph.D. and Richard H. Becker, Ph.D.
Department of Earth, Ecological & Environmental Sciences
The University of Toledo

Kendall Hunt
p u b l i s h i n g   c o m p a n y

Cover image © Shutterstock, Inc.

Kendall Hunt
publishing company

www.kendallhunt.com
Send all inquiries to:
4050 Westmark Drive
Dubuque, IA 52004-1840

Printed in the United States of America
10 9 8 7 6 5 4 3

Contents

Note to Students

The commercially published laboratory manuals for introductory physical geology that are sold nationally are typically beautifully illustrated with color photographs, drawings, charts and maps. They also generally have in-depth background discussions of all the topics dealt with. These manuals, however, have prices in the $60–80 range, which is too much for what is usually a one-hour course, especially since the manuals cannot be sold back to bookstores because some exercises require pages to be removed. It was to save students money that this present manual was written. If you have become accustomed to colorful, glitzy books with copious texts and illustrations then you might find this one a bit brief and dull in comparison. It does, however, have all the same kinds of exercises as the more expensive manuals and this, ultimately, is the most important thing.

This manual is intended to be a stand-alone textbook and was written specifically for the Introductory Physical Geology Laboratory (EEES 1020) at the University of Toledo. The vast majority of students in this course will also be taking the Physical Geology lecture course (EEES 1010) and so will have another, more comprehensive textbook. Nearly everything dealt with in the laboratory will also be covered to some extent in the lecture. The student will, therefore, find it helpful to refer to the appropriate pages in their lecture textbook for supplementary background readings on the topics covered in this laboratory manual.

The instructor will supply students with rock and minerals specimens, and certain other supplies. There are, however, other supplies that students need to provide for themselves and these are as follows: (1) required supplies – pencil with red or other colored lead; and (2) optional but strongly recommended supplies – magnifying glass, electronic calculator, and long (~30 cm) metric-scale ruler.

About the Author

Dr. James A. Harrell is a Professor of Geology in the Department of Environmental Sciences at the University of Toledo. For nearly three decades he has taught a variety of courses, including in recent years: Physical Geology, Geologic Hazards, Megascopic Petrology, Microscopic Petrology, Multivariate Geostatistics, and Archaeology of Ancient Egypt. His research is on the archaeological geology of ancient Egypt with particular emphasis on the varieties, sources, quarrying, and applications of the rocks and minerals used by the ancient Egyptians. Dr. Richard Becker is an Assistant Professor in the Department of Environmental Sciences at the University of Toledo. He uses an interdisciplinary research approach (remote sensing, geochemistry, hydrologic modeling, and field techniques) to investigate a wide range of geological and environmental problems including the the assessment of responses of natural systems to climatic and human activities. These have included studying the impacts on the Nubian aquifer in Northeastern Africa and Nile watershed of the construction of the Aswan High Dam ranging from changing recharge in the system to changing subsidence in the Nile Delta, as well as investigating the effect climate change has had on this system; studying the role land use has had on water quality and mapping potential harmful algal blooms in the Great Lakes.

Chapter 1

Minerals

Note: *it is recommended that you bring a magnifying glass to class.*

Elements to Minerals

All the solid, liquid and gaseous materials making up the Earth are composed of **atoms**. Atoms are the smallest division of matter obtainable by chemical means. Every atom consists of a compact **nucleus** containing positively charged **protons** and uncharged **neutrons**, and surrounding the nucleus are orbiting clouds of negatively charged **electrons**. Atoms are distinguished from one another by the number of protons in the nucleus, and it is by this, the so-called **atomic number**, that the **elements** are defined. The elements are, thus, the different varieties of atoms.

All atoms having the same number of protons belong to the same element. There are 88 elements that occur naturally in the Earth's crust (another

TABLE 1.1	
Most Abundant Elements in the Earth's Crust	
Element*	**Percentage of Crust by Weight**
oxygen (^{8}O)	46.60
silicon (^{14}Si)	27.72
aluminum (^{13}Al)	8.13
iron (^{26}Fe)	5.00
calcium (^{20}Ca)	3.63
sodium (^{11}Na)	2.83
potassium (^{19}K)	2.59
magnesium (^{12}Mg)	2.09
	98.59

*Given within parentheses are the standard symbol and atomic number for each element.

20 or so have been created by scientists under laboratory conditions). Interestingly, two of these elements, oxygen and silicon, make up about 74 percent of the crust by weight (see Table 1.1). The addition of aluminum brings the total weight percent to 82, and another five elements (iron, calcium, sodium, potassium and magnesium) increases it to about 99 percent. The other 80 elements, thus, make up only about 1 percent of the Earth's crust.

The elements combine to form the solid materials that we call **minerals**.

Mineral Defined

A mineral is a **solid** material that exhibits the following two properties: (1) it has a **crystalline structure** [that is, an orderly and constant three-dimensional arrangement of the constituent atoms]; and (2) it has a **fixed elemental composition** [that is, the types and relative proportions of the constituent elements are constant and can be represented by a **chemical formula**]. To these defining properties two less critical ones may be added; namely that minerals are **naturally occurring** and **inorganic**. Thus, by these last two requirements, crystalline solids produced by synthetic means in laboratories or by organisms in nature are, technically, not minerals.

There are thousands of known minerals but only a small number of them (the so-called **rock-forming minerals**) make up the bulk of the Earth's crust. Not surprisingly, these minerals consist largely of the eight most abundant elements, and especially oxygen and silicon. When the latter two elements occur together the material is said to contain **silica** (SiO_2) and is referred to as a **silicate** mineral. The silicates are the most important and abundant of the rock-forming minerals.

Gems (or gemstones) are minerals that are especially beautiful and durable, and, thus, suitable for jewelry. Their rarity and other properties determine whether they are considered **precious** or **semi-precious** stones.

Mineral Properties and Identification

Minerals are identified by their physical and other properties.

Color

Color is a mineral's most obvious property but it is not always a reliable one. This is so because some minerals may occur in a variety of colors due to the presence of small amounts of impurities (that is, elements not included in the mineral's chemical formula). In general, however, a given mineral will exhibit consistently light or dark colors.

Luster

Luster refers to the appearance of a mineral's surface in reflected light. It is independent of its color. The characterization of luster is somewhat subjective and is usually described in the following terms: **vitreous** (bright and shiny like glass), **adamantine** (brilliantly bright like a diamond), **greasy** (looks like it's covered with a thin layer of oil), **resinous** (looks like resin), **silky** (looks like silk), **pearly** (looks like iridescent pearls), and **dull and earthy** (not reflective at all). These are the **non-metallic** lusters, but minerals can also have **metallic** (looks like a metal) or **sub-metallic** (transitional between metallic and nonmetallic) lusters.

Hardness

A mineral's **hardness** is its resistance to scratching. Geologists use the **Mohs hardness scale** to characterize the relative scratch resistance of minerals (see Table 1.2). Hardness is determined by either scratching a mineral with a material of known hardness or vice versa. These materials will be provided in class.

TABLE 1.2		
Mohs Hardness Scale		
Mohs Hardness	**Index Minerals**	**Hardness of Common Materials**
1	talc	
2	gypsum	fingernail (2.2)
3	calcite	copper penny (3.1)
4	fluorite	
5	apatite	steel nail or knife blade (5.1–5.3)
6	orthoclase (K-feldspar)	glass plate (5.5) steel file (6.5)
7	quartz	porcelain streak plate (7.0)
8	topaz	
9	corundum	
10	diamond	

Cleavage and Fracture

Cleavage is the tendency of some minerals to break consistently and repeatedly in certain directions. These are the "planes of weakness" within a mineral and are an outward manifestation of its internal crystalline structure. Not every mineral exhibits cleavage but those that do will have one of the following types: **basal** (one direction only), **prismatic** (two directions at either right angles [90°] or NOT at right angles to each other), **cubic** (three directions at right angles to each other), **rhombohedral** (three directions NOT at right angles to each other), **octahedral** (four directions NOT at right angles to each other), and **dodecahedral** (six directions NOT at right angles to each other). These types of cleavage are illustrated in Figure 1.1. In this figure, hypothetical mineral fragments are shown where all the visible surfaces are produced by breakage along cleavage

FIGURE 1.1 Types of Cleavage in Minerals

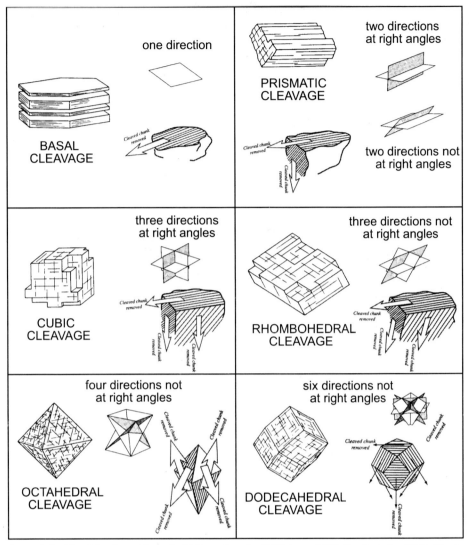

Art on left side of fox from Manual of Mineralogy by C. Klein and C.S. Hurlbut, Jr., 1993. Used by permission of John Wiley & Sons, Inc.

Art in the upper right side of each box from Physical Geology: Laorabory Text and Manual by David Dallmeyer. Copyright © 2000 by Kendall Hunt Publishing Company. Reprinted by permission.

Art in the lower right side of each box from Physical Geology Laboratory Manual, 4th edition by Karen M. Woods. Copyright 2006 by Kendall Hunt Publishing Company. Reprinted by permission.

planes (as indicated by the sets of parallel lines within the fragments).

Cleavage can be a difficult property to recognize. This is true both because not all the cleavage directions in a given mineral are equally well developed and because not all mineral surfaces will clearly show cleavage. To identify the type of cleavage in a mineral that possesses this property, you need to closely examine all the surfaces of a sample, preferably under a strong light and with a magnifying glass. Under these viewing conditions the flat cleavage surfaces, which tend to be highly reflective, will sparkle when the sample is turned in the light. If you are having trouble seeing the cleavage, try using a binocular microscope if one is available in the classroom.

A few minerals exhibit a "false basal cleavage" known as **parting**. This type of breakage does not result from the internal crystalline structure but rather is caused by other factors. It is sometimes seen in crystalline masses that are not fragments of individual mineral grains but are instead aggregates of many grains. These masses may be either "microcrystalline" (grains not visible to the unaided eye) or "macrocrystalline" (grains easily visible to the unaided eye).

Minerals lacking cleavage altogether are said to exhibit **fracture**. The fracture surfaces may be described as either **conchoidal**, if smooth and curved with concentric ridges, or **uneven**, if rough and irregular.

Diaphaneity

The ease with which light passes through a mineral is its **diaphaneity**. This property is best seen along the thin edges of samples held up to a bright light. A mineral may, accordingly, be described as **transparent** (most or all the light passes through and objects viewed through the sample are clearly seen), **translucent** (little light passes through and objects viewed through the sample are obscured), and **opaque** (no light passes through).

Streak

If a mineral is dragged across a porcelain plate and its hardness is less than that of the plate, a powdery residue will be left behind. The color of this residue is the mineral's **streak**. Interestingly, the streak color is not necessarily the same as the mineral color. The latter is affected by impurities whereas the streak is not. The streak is, thus, a fairly reliable diagnostic property of minerals, especially those with a metallic luster.

Specific Gravity

The **specific gravity** of a mineral is the ratio of its weight to that of an equal volume of water. Under classroom conditions, specific gravity has the same numerical value as density which is the weight of the sample divided by its volume. The former is dimensionless whereas the latter has units of grams per cubic centimeter. The precise determination of specific gravity requires special equipment but you can get a general idea of relative magnitudes by simply hefting a mineral sample in your hand. By comparing the weights of two samples of similar size, one a known mineral and the other an unknown, you can roughly estimate the latter's specific gravity.

Crystal Form

When minerals grow into an open space, such as a cavity inside a rock, they will usually develop into distinctive forms called **crystals**. Like cleavage, crystal form is an external expression of the internal crystalline structure. Crystals will exhibit a number of faces or sides. Interestingly, however, the cleavage directions and the orientations of the crystal faces are not necessarily the same. Also, minerals that lack cleavage can still form crystals. A word of caution: crystal faces may be mistaken for cleavage surfaces and vice versa. Most of the samples you will be working with in class do not show crystal forms either because they are broken fragments of crystals or because they come from an aggregate of intergrown grains where crystals never had a chance to form.

Other Properties

Minerals exhibit many other properties, and a few of these are useful for identifying some of the minerals you will be working with in class. **Magnetism** refers to the attraction of a mineral to a magnet. Some iron-rich minerals are noticeably magnetic. **Reaction to acid** is characteristic of some carbonate-rich (CO_3) minerals. To test a sample apply ONE SMALL DROP of dilute (5–10 percent) hydrochloric acid. A reaction is indicated if effervescence (the fizz caused by the release of CO_2 gas) is observed. Some minerals have a distinctive **feel**, **odor**, or **taste**. Others may exhibit fine **striations** (due to crystal twinning) or **perthite lamellae** (due to minute mineral inclusions) on cleavage surfaces.

Exercise 1

Your instructor will provide you with a set of 22 mineral specimens as well as supplies for testing their properties (copper penny, steel nail, glass plate, porcelain streak plate, magnet, and dilute acid). These specimens include all the important rock-forming minerals. Carefully examine each specimen and neatly record its properties on a "Mineral Data Sheet." Use the "Mineral Identification Key" in Table 1.3 to determine the mineral name. This table includes all the minerals in your class set plus a few others that are not.

A useful way to start the identification process is to first organize the minerals into three groups: (1) non-metallic luster and light colors, (2) non-metallic luster and dark colors, and (3) sub-metallic or metallic luster. Label each pile of specimens with a small piece of paper. Further subdivide the minerals in groups 1 and 2 according to whether they are harder or softer than glass, and label accordingly. At this point in the exercise you have worked your way through the first two levels of the identification key. Following this key, continue subdividing your minerals on the basis of their other properties until you have them all identified.

Hand in the filled-out mineral data sheets to your instructor. These will be graded and returned to you.

TABLE 1.3
Mineral Identification Key

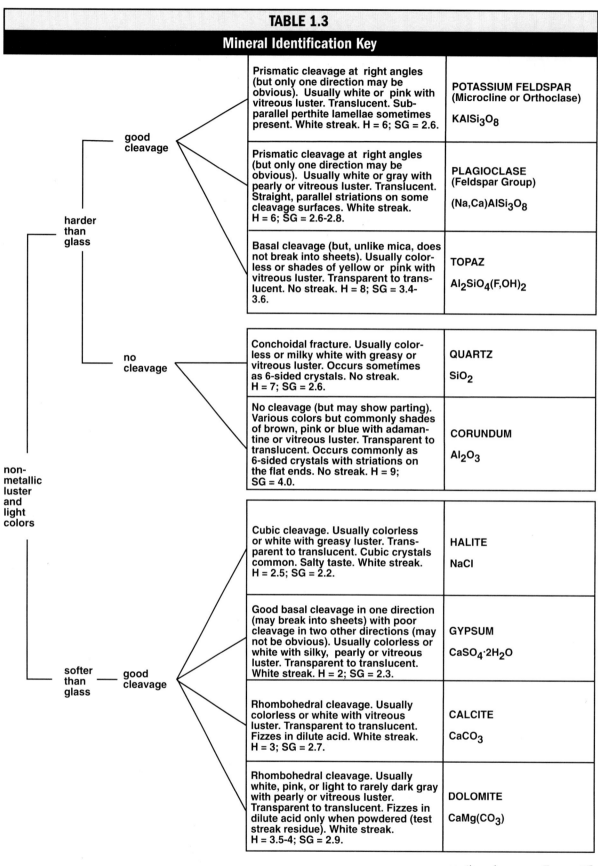

harder than glass — **good cleavage**	Prismatic cleavage at right angles (but only one direction may be obvious). Usually white or pink with vitreous luster. Translucent. Sub-parallel perthite lamellae sometimes present. White streak. H = 6; SG = 2.6.	**POTASSIUM FELDSPAR** (Microcline or Orthoclase) $KAlSi_3O_8$
	Prismatic cleavage at right angles (but only one direction may be obvious). Usually white or gray with pearly or vitreous luster. Translucent. Straight, parallel striations on some cleavage surfaces. White streak. H = 6; SG = 2.6-2.8.	**PLAGIOCLASE** (Feldspar Group) $(Na,Ca)AlSi_3O_8$
	Basal cleavage (but, unlike mica, does not break into sheets). Usually colorless or shades of yellow or pink with vitreous luster. Transparent to translucent. No streak. H = 8; SG = 3.4-3.6.	**TOPAZ** $Al_2SiO_4(F,OH)_2$
no cleavage	Conchoidal fracture. Usually colorless or milky white with greasy or vitreous luster. Occurs sometimes as 6-sided crystals. No streak. H = 7; SG = 2.6.	**QUARTZ** SiO_2
	No cleavage (but may show parting). Various colors but commonly shades of brown, pink or blue with adamantine or vitreous luster. Transparent to translucent. Occurs commonly as 6-sided crystals with striations on the flat ends. No streak. H = 9; SG = 4.0.	**CORUNDUM** Al_2O_3
softer than glass — **good cleavage**	Cubic cleavage. Usually colorless or white with greasy luster. Transparent to translucent. Cubic crystals common. Salty taste. White streak. H = 2.5; SG = 2.2.	**HALITE** $NaCl$
	Good basal cleavage in one direction (may break into sheets) with poor cleavage in two other directions (may not be obvious). Usually colorless or white with silky, pearly or vitreous luster. Transparent to translucent. White streak. H = 2; SG = 2.3.	**GYPSUM** $CaSO_4 \cdot 2H_2O$
	Rhombohedral cleavage. Usually colorless or white with vitreous luster. Transparent to translucent. Fizzes in dilute acid. White streak. H = 3; SG = 2.7.	**CALCITE** $CaCO_3$
	Rhombohedral cleavage. Usually white, pink, or light to rarely dark gray with pearly or vitreous luster. Transparent to translucent. Fizzes in dilute acid only when powdered (test streak residue). White streak. H = 3.5-4; SG = 2.9.	**DOLOMITE** $CaMg(CO_3)$

non-metallic luster and light colors

continued on page 7

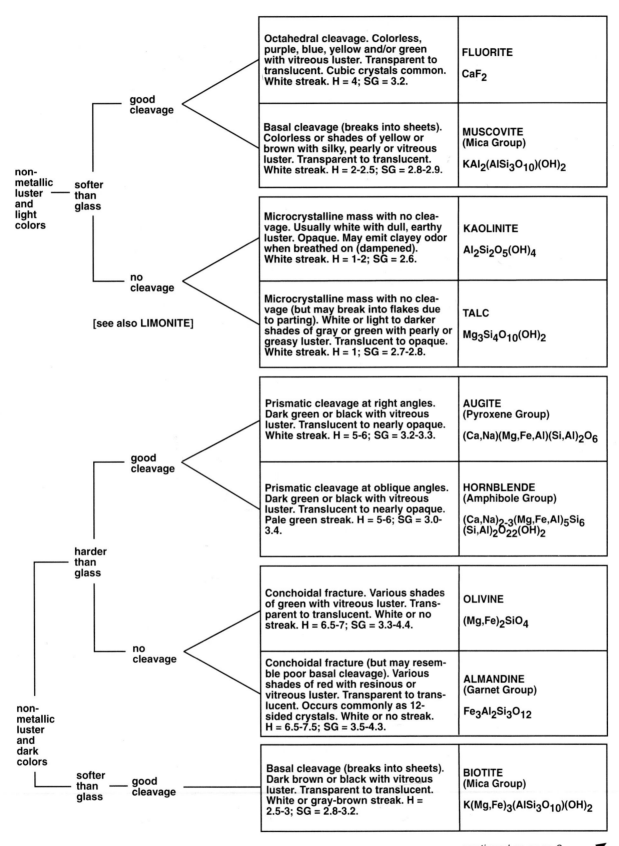

non-metallic luster and light colors — **softer than glass**

good cleavage

Octahedral cleavage. Colorless, purple, blue, yellow and/or green with vitreous luster. Transparent to translucent. Cubic crystals common. White streak. H = 4; SG = 3.2.

FLUORITE

CaF_2

Basal cleavage (breaks into sheets). Colorless or shades of yellow or brown with silky, pearly or vitreous luster. Transparent to translucent. White streak. H = 2-2.5; SG = 2.8-2.9.

MUSCOVITE
(Mica Group)

$KAl_2(AlSi_3O_{10})(OH)_2$

no cleavage

Microcrystalline mass with no cleavage. Usually white with dull, earthy luster. Opaque. May emit clayey odor when breathed on (dampened). White streak. H = 1-2; SG = 2.6.

KAOLINITE

$Al_2Si_2O_5(OH)_4$

Microcrystalline mass with no cleavage (but may break into flakes due to parting). White or light to darker shades of gray or green with pearly or greasy luster. Translucent to opaque. White streak. H = 1; SG = 2.7-2.8.

TALC

$Mg_3Si_4O_{10}(OH)_2$

[see also LIMONITE]

harder than glass

good cleavage

Prismatic cleavage at right angles. Dark green or black with vitreous luster. Translucent to nearly opaque. White streak. H = 5-6; SG = 3.2-3.3.

AUGITE
(Pyroxene Group)

$(Ca,Na)(Mg,Fe,Al)(Si,Al)_2O_6$

Prismatic cleavage at oblique angles. Dark green or black with vitreous luster. Translucent to nearly opaque. Pale green streak. H = 5-6; SG = 3.0-3.4.

HORNBLENDE
(Amphibole Group)

$(Ca,Na)_{2-3}(Mg,Fe,Al)_5Si_6(Si,Al)_2O_{22}(OH)_2$

no cleavage

Conchoidal fracture. Various shades of green with vitreous luster. Transparent to translucent. White or no streak. H = 6.5-7; SG = 3.3-4.4.

OLIVINE

$(Mg,Fe)_2SiO_4$

Conchoidal fracture (but may resemble poor basal cleavage). Various shades of red with resinous or vitreous luster. Transparent to translucent. Occurs commonly as 12-sided crystals. White or no streak. H = 6.5-7.5; SG = 3.5-4.3.

ALMANDINE
(Garnet Group)

$Fe_3Al_2Si_3O_{12}$

non-metallic luster and dark colors — **softer than glass** — **good cleavage**

Basal cleavage (breaks into sheets). Dark brown or black with vitreous luster. Transparent to translucent. White or gray-brown streak. H = 2.5-3; SG = 2.8-3.2.

BIOTITE
(Mica Group)

$K(Mg,Fe)_3(AlSi_3O_{10})(OH)_2$

continued on page 8

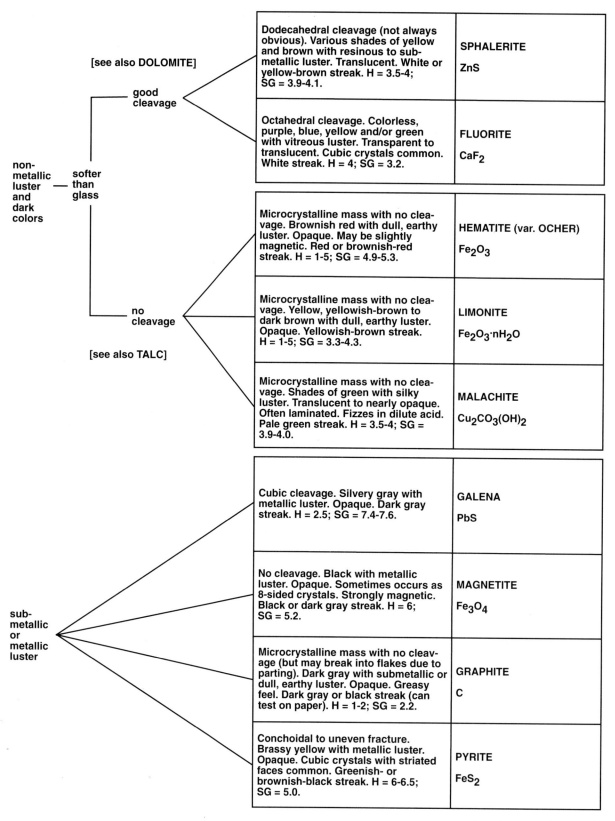

	Dodecahedral cleavage (not always obvious). Various shades of yellow and brown with resinous to sub-metallic luster. Translucent. White or yellow-brown streak. H = 3.5-4; SG = 3.9-4.1.	SPHALERITE ZnS
	Octahedral cleavage. Colorless, purple, blue, yellow and/or green with vitreous luster. Transparent to translucent. Cubic crystals common. White streak. H = 4; SG = 3.2.	FLUORITE CaF$_2$
	Microcrystalline mass with no clea-vage. Brownish red with dull, earthy luster. Opaque. May be slightly magnetic. Red or brownish-red streak. H = 1-5; SG = 4.9-5.3.	HEMATITE (var. OCHER) Fe$_2$O$_3$
	Microcrystalline mass with no clea-vage. Yellow, yellowish-brown to dark brown with dull, earthy luster. Opaque. Yellowish-brown streak. H = 1-5; SG = 3.3-4.3.	LIMONITE Fe$_2$O$_3 \cdot$nH$_2$O
	Microcrystalline mass with no clea-vage. Shades of green with silky luster. Translucent to nearly opaque. Often laminated. Fizzes in dilute acid. Pale green streak. H = 3.5-4; SG = 3.9-4.0.	MALACHITE Cu$_2$CO$_3$(OH)$_2$
	Cubic cleavage. Silvery gray with metallic luster. Opaque. Dark gray streak. H = 2.5; SG = 7.4-7.6.	GALENA PbS
	No cleavage. Black with metallic luster. Opaque. Sometimes occurs as 8-sided crystals. Strongly magnetic. Black or dark gray streak. H = 6; SG = 5.2.	MAGNETITE Fe$_3$O$_4$
	Microcrystalline mass with no cleav-age (but may break into flakes due to parting). Dark gray with submetallic or dull, earthy luster. Opaque. Greasy feel. Dark gray or black streak (can test on paper). H = 1-2; SG = 2.2.	GRAPHITE C
	Conchoidal to uneven fracture. Brassy yellow with metallic luster. Opaque. Cubic crystals with striated faces common. Greenish- or brownish-black streak. H = 6-6.5; SG = 5.0.	PYRITE FeS$_2$

non-metallic luster and dark colors — softer than glass

good cleavage [see also DOLOMITE]

no cleavage [see also TALC]

sub-metallic or metallic luster

continued on page 9

sub-metallic or metallic luster	Micro- or macrocrystalline mass with no cleavage. Brassy yellow with metallic luster and often purplish iridescent tarnish. Opaque. Greenish-black streak. H = 3.5-4; SG = 4.1-4.3.	**CHALCOPYRITE** $CuFeS_3$
	Uneven fracture. Silvery gray with metallic luster or red with submetallic luster. Opaque. Platy crystals common. May be slightly magnetic. Red or brownish-red streak. H = 5.5-6.5; SG = 5.3.	**HEMATITE** **(var. SPECULARITE)** Fe_2O_3

A NOTE REGARDING MINERAL COLORS

"LIGHT COLORS" are considered to be: colorless, white, yellow, pink, and light to intermediate shades of all other colors.

"DARK COLORS" are considered to be: black, and dark shades of all other colors.

Mineral Data Sheet

Student's Name _____

Sample Number	Luster	Hardness	Cleavage or Fracture	Overall Color	Streak Color	Other Properties	MINERAL NAME

Igneous Rocks

Note: *it is recommended that you bring a magnifying glass and metric-scale ruler to class.*

The Rock Cycle

When grains of one or more minerals are combined to form a solid aggregate we say that such a material is a **rock**. There are three categories of rock: **igneous**, **metamorphic**, and **sedimentary**. Generally speaking, igneous rocks are formed through the cooling and solidification of magma (molten or liquid rock). Sedimentary rocks form from the compaction and cementation of sediments produced, in large part, by the weathering and erosion of pre-existing rocks. And metamorphic rocks form when pre-existing rocks are altered by high temperatures and pressures. The interrelationships among the three rock types are embodied within what has been called the "Rock Cycle" (see Figure 2.1).

The current chapter is concerned only with igneous rocks, and subsequent chapters will deal with those of sedimentary and metamorphic origins.

Origin of Igneous Rocks

Igneous rocks form through the crystallization of magma (that is, the elements combine to form minerals). If crystallization occurs below the Earth's surface, the resulting rocks are said to be **plutonic** or **intrusive**. The latter term derives from the fact that magma always moves upwards through the crust and so "intrudes" the pre-existing rocks. Intrusive igneous rock bodies have a variety of names that reflect their size and geometry (see Figure 2.2). Sometimes magma works its way to the Earth's surface where it is then erupted through a **volcano**. When extruded onto the surface, magma is referred to as **lava**. Magma, along with pieces of solid rock, can be blasted out of volcanoes during explosive eruptions and such materials, which eventually fall to the surface, are called **pyroclastics**. The so-called **volcanic** or **extrusive** igneous rocks are those formed from pyroclastic deposits and lava flows.

Pyroclastic rocks, it should be noted, do not entirely conform to the simple definition of igneous rocks given above ("crystallization from magma"). The individual grains (**pyroclasts**) are certainly formed from crystallized magma but the pyroclastic rock itself is an aggregate of grains that has been hardened into a solid mass (**lithified**) through compaction and cementation (just as in sedimentary rocks).

Igneous Rock Properties and Classification

Igneous rocks are named by applying a classification scheme like that in Table 2.1. To use it you need to recognize two properties of the rocks: the **composition** (that is, the minerals present and their relative abundances), and the **texture** (that is, the size of the constituent mineral grains, the way the grains are joined, and other physical characteristics).

The minerals present in igneous rocks are all ones that you worked with in Exercise 1. They tend to occur in four general groupings referred to as **felsic**, **intermediate**, **mafic**, and **ultramafic** (see Table 2.1). Identifying the minerals in rocks is more difficult than when working with large, relatively pure mineral samples as in the previous exercise, and is nearly impossible (except when using sophisticated equipment) when the

FIGURE 2.1 Types of Igneous Deposits

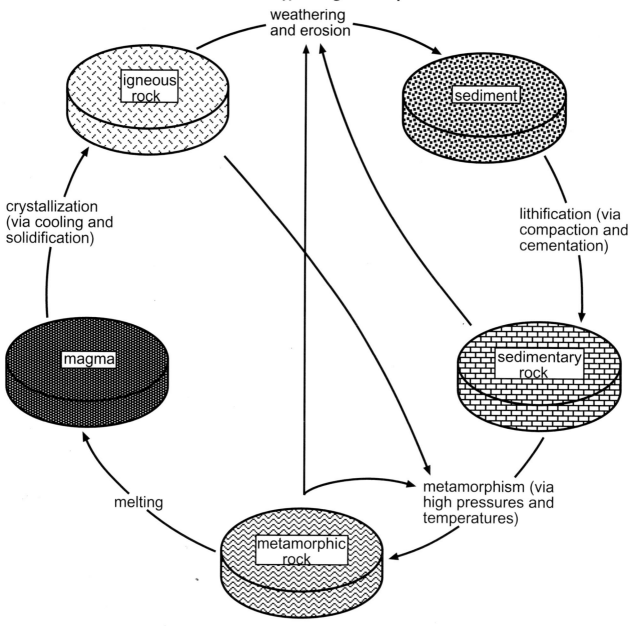

individual grains are too small to see. In such cases the rock color will give an indication of the general composition. Do not, however, rely only on color because, as in some minerals, a given type of igneous rock can come in a range of colors due to impurities. Try to identify the minerals in each rock. A magnifying glass and good lighting will make it easier, and you can also use a binocular microscope if one is available.

Most igneous rocks have a **crystalline texture** (that is, they consist of interlocking grains that grew together during crystallization from a magma). If essentially all the individual grains are large enough to be clearly seen with the unaided eye (or even with a magnifying glass), the texture is said to be **phaneritic**, and if the grains are too small to be seen the rock has an **aphanitic** texture. Phaneritic rocks with exceptionally large grains

FIGURE 2.2 The Rock Cycle

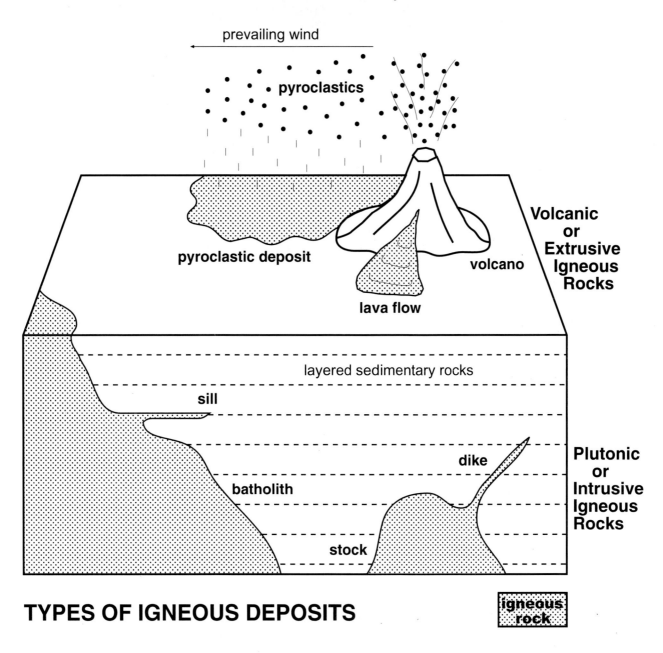

prevailing wind

pyroclastics

pyroclastic deposit

volcano

Volcanic or Extrusive Igneous Rocks

lava flow

layered sedimentary rocks

sill

dike

batholith

Plutonic or Intrusive Igneous Rocks

stock

TYPES OF IGNEOUS DEPOSITS

igneous rock

exhibit a **pegmatitic** texture. In these rocks the grains are at least a few centimeters across and can even be up to several meters in dimension! Some rocks will have grains of two widely different sizes. These have phaneritic crystals (called **phenocrysts**) set in an aphanitic **matrix**. Such rocks are called **porphyry** if the phenocrysts make up *over 20 percent* of the total surface area. On the right side of Table 2.1 is an illustration labeled

"20% comparator." This simulates the appearance of a rock with an aphanitic matrix and phenocrysts (the large black rectangles) that constitute exactly 20 percent of the area within the square.

The **fragmental textures** of pyroclastic rocks are essentially the same as those found in many sedimentary rocks. The pyroclasts, which are rock and mineral fragments, are not intergrown as in a crystalline texture, but rather are compacted and

Chapter 2 **Igneous Rocks**   19

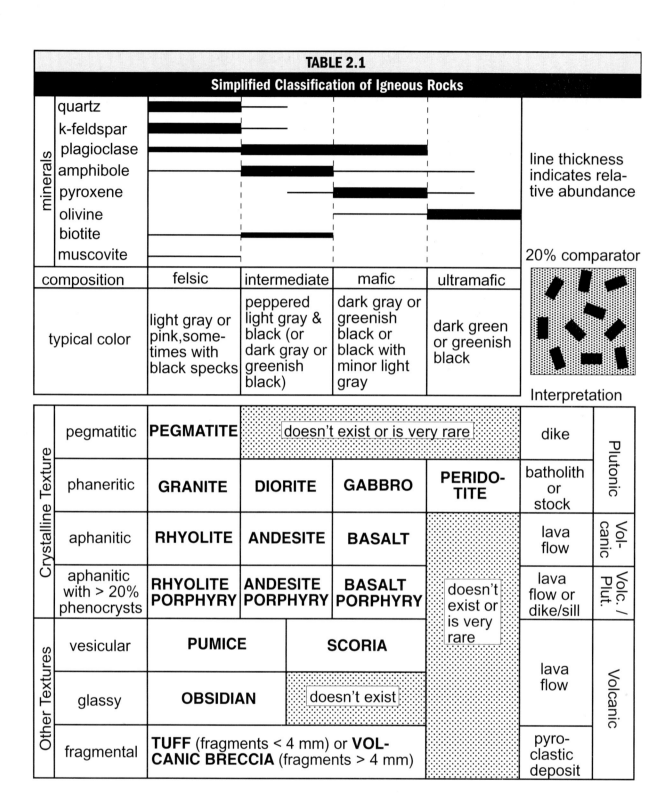

TABLE 2.1

Simplified Classification of Igneous Rocks

minerals					line thickness indicates relative abundance
quartz					
k-feldspar					
plagioclase					
amphibole					
pyroxene					
olivine					
biotite					
muscovite					20% comparator

composition	felsic	intermediate	mafic	ultramafic	
typical color	light gray or pink, sometimes with black specks	peppered light gray & black (or dark gray or greenish black)	dark gray or greenish black or black with minor light gray	dark green or greenish black	Interpretation

		felsic	intermediate	mafic	ultramafic		
Crystalline Texture	pegmatitic	**PEGMATITE**	doesn't exist or is very rare			dike	Plutonic
	phaneritic	**GRANITE**	**DIORITE**	**GABBRO**	**PERIDO-TITE**	batholith or stock	
	aphanitic	**RHYOLITE**	**ANDESITE**	**BASALT**	doesn't exist or is very rare	lava flow	Volcanic
	aphanitic with > 20% phenocrysts	**RHYOLITE PORPHYRY**	**ANDESITE PORPHYRY**	**BASALT PORPHYRY**		lava flow or dike/sill	Volc. / Plut.
Other Textures	vesicular	**PUMICE**	**SCORIA**			lava flow	Volcanic
	glassy	**OBSIDIAN**	doesn't exist				
	fragmental	**TUFF** (fragments < 4 mm) or **VOL-CANIC BRECCIA** (fragments > 4 mm)				pyro-clastic deposit	

cemented together. The names applied to these rocks are based only on the predominant size of the pyroclasts.

Lava on the Earth's surface sometimes cools so rapidly that minerals do not have time to crystallize. The resulting rock, **obsidian**, is a kind of glass and so has a **glassy texture**. This is the same process by which window glass is made: grains of quartz (plus some additives) are melted and then rapidly cooled. The resulting glassy solid is no longer quartz because it lacks a crystalline tructure, one of the key properties of a mineral.

The last igneous texture, **vesicular**, describes thoe rocks that are full of holes ("vesicles") and look a it like Swiss cheese. **Pumice** tends to be white in color and has a super abundance of very small vesicles whereas **scoria** is black, brown or red, and has fewer and larger vesicles than pumice. Pumice is basically just vesicular obsidian, and scoria may consist partially of glass but is mostly crystalline rock. The vesicles in both rocks are formed by gas bubbles trapped in the rapidly cooling lava. The same phenomenon occurs in the frothy head on an overly warmed or agitated Coca-Cola.

The textures of igneous rocks are closely related to the different types of igneous deposits shown in Figure 2.2. These relationships are indicated in Table 2.1 under "Interpretation."

Exercise 2

Your instructor will provide you with a set of 11 igneous rock specimens. Carefully examine each specimen and <u>neatly</u> record its properties on an "Igneous Rock Data Sheet." Use the classification in Table 2.1 to determine the rock name. This table includes all the different rock types in your set (but the set does not necessarily include all the rock types in the table!).

To start the identification process, subdivide the specimens into two groups: (1) crystalline textures, and (2) other textures [vesicular, glassy and fragmental]. Next, using a combination of mineralogy and color, assign names to each specimen.

Hand in the filled-out data sheets to your instructor. These will be graded and returned to you.

Igneous Rock Data Sheet

Student's Name _____

Sample Number	Texture	Color	Mineralogy	Other Properties	ROCK NAME

Chapter 3

Sedimentary Rocks

Note: *it is recommended that you bring a magnifying glass and metric-scale ruler to class.*

Weathering and Sediments

When rocks in the Earth's crust are exposed at the surface they begin to disintegrate. The disintegration process is referred to as **weathering**, and the rock and mineral fragments produced are called **sediment**. There are two categories of weathering: mechanical and chemical.

Mechanical weathering en-compasses those processes that can physically break a rock into fragments. It includes freezing and thawing of water in cracks ("frost wedging", a common wintertime problem for streets in the Upper Midwest), thermal expansion and contraction during the daily heating cycle of sunlight and darkness, burrowing activities of animals, growth of plant roots into cracks, and wear-and-tear ("abrasion") on rock fragments transported by water or other agents.

Chemical weathering, as the name implies, disintegrates rocks by chemical rather than physical means. The processes include "dissolution" (usually by slightly acidic water), "oxidation" (a reaction with oxygen in air and water), and "hydrolysis" (a reaction with water).

Some of the sediment produced by weathering on the continental surfaces is carried to the ocean by rivers. However, not all the sediment accumulating on the ocean bottoms is of this origin. Great quantities of sediment are derived from the skeletons of plants (algae) and animals (corals, clams, and other invertebrates) that live in the sea. Other marine sediments, in contrast, have a non-organic origin and are the result of chemical reactions occurring when seawater evaporates (as it sometimes does when small seas become geographically isolated from the world ocean).

Sediment is picked up (**eroded**) by winds, glaciers or rivers on the continents, and by currents or waves in the oceans, and then **transported** many miles (sometimes hundreds) before being permanently **deposited**. The places of deposition are the so-called **depositional environments** for sedimentary rocks (see Table 3.1).

The vast majority of sediment grains are either **mineral fragments** (pieces of a single mineral), **rock fragments** (pieces of an aggregate of two or more minerals), or **shell fragments** (pieces of animal shells or skeletons).

Lithification

Sediments become sedimentary rocks through the processes of **compaction** and **cementation**. Compaction results from the weight of overlying sediments which, for a given sediment layer, increases over time as deposition continues and the layer becomes more deeply buried. The same effect is achieved, for example, by trash compactors that can produce solid blocks of paper and cans.

The pore spaces between sediment grains are often filled with water. The water is not pure but rather has a number of "dissolved minerals" in it. Under certain circumstances, these minerals can **precipitate** (that is, come out of solution) in the pores and so act as a cement or glue that holds the grains together. The same process produces the white, crusty deposits that form inside your teakettle and water pipes (the "harder" the water, the more deposits). Several minerals can form cements in sedimentary rocks and the most common of these are calcite, quartz, and hematite.

TABLE 3.1

Simplified Classification of Sedimentary Rocks

General Texture	Specific Texture		Composition	ROCK NAME	Principal Depositional Environment(s)
clastic	grains mostly larger than 2 mm (i.e., pebbles and cobbles)	grains rounded	rock fragments of any type (can have a non-silicic composition) plus cement; any color possible	conglomerate	gravel deposited on/in alluvial fan, braided river, or beach
		grains angular		breccia	
	grains mostly between 0.0625 and 2 mm (i.e., sand)	*mainly quartz, calcite or hematite cement*	grains mostly quartz; white, shades of gray, brown or yellow	quartz sandstone	sand deposited on/in meandering or braided river, eolian dune, beach, river delta or nearshore shallow marine
			grains mostly feldspar; pinkish or reddish	arkosic sandstone	
			grains mostly rock fragments; shades of brown or gray with dark speckles	lithic sandstone	
			grains of any composition but with clay matrix; dark gray or green	graywacke sandstone	sand deposited by deep marine turbidite
	grains mostly less than 0.0625 mm (i.e., silt and clay)	feels gritty, commonly not laminated	finer-grained version of cemented sandstone; shades of gray, brown or green	siltstone	mud deposited in/on meandering river, river delta, lake bottom, or shallow to deep marine
		feels smooth, commonly laminated	mostly clay minerals; shades of gray or green	shale	
	no set grain size but typically very fine-grained; soft and porous		microscopic opaline diatom shells; white or light gray	diatomite	deep marine
crystalline	aphanitic; hard (scratches glass) and dense		quartz; shades of gray with reddish or yellowish stains	chert	deep marine or re-placement of limestone

General Texture	Principal Mineral	Specific Texture and Composition		ROCK NAME	Principal Depositional Environment(s)
clastic	calcite	shades of gray, sometimes with yellow or pink stains	contains conspicuous shell or other skeletal fragments — rock consists entirely of shell fragments that are cemented together	coquina or coquinoid limestone	shallow marine
			contains conspicuous shell or other skeletal fragments — rock contains some shell fragments in a matrix of fine-grained carbonate sediment	fossiliferous limestone	shallow marine
			fine-grained and soft (can be cut with a fingernail)	chalk	deep marine
			fine-grained and hard (cannot be cut with a fingernail)	micrite or micritic limestone	shallow or deep marine
			contains mostly spherical, sand-size grains (ooliths)	oolite or oolitic limestone	shallow marine
			aphanitic to phaneritic texture	crystalline limestone	recrystallized limestone
crystalline	dolomite	aphanitic to phaneritic texture; light gray to brownish gray		dolostone	shallow marine; recrystallized limestone
	gypsum	aphanitic to phaneritic texture; white, light gray or pink		rock gypsum	deep marine; sea-water evaporation
	halite	phaneritic; colorless, white or light gray		rock salt	deep marine; sea-water evaporation
	hematite or limonite	aphanitic texture; shades of red, yellow, orange or brown		ironstone	continental; replacement by iron oxides
clastic or amorphous	bitumen	aphanitic texture with or without plant fragments; black		coal (bituminous)	swamp (usually deltaic or riverine)

(limestones)

Sedimentary Rock Properties and Classification

A classification scheme like that in Table 3.1 is used to name sedimentary rocks. As with igneous rocks, the classification is based on both **composition** and **texture**. The first distinction to be made is between **silicic** and **non-silicic** rocks. As the name implies, silicic rocks consist mostly of silicate minerals (mainly quartz, feldspar, and clay) whereas the non-silicic rocks are composed of mostly of non-silicate minerals (such as calcite, dolomite, gypsum, and halite).

Although the compositions are very different, the two rock groups are not always easy to distinguish. Most of the non-silicic rocks are carbonates (**dolostones** and especially **limestones**), and so will effervesce when tested with acid. Care must be taken in applying the acid test, however, because a calcite cement in a **sandstone** will also effervesce. You, thus, need to look closely at what part of the rock is reacting to the acid: essentially all of it in carbonates (keeping in mind that the dolostone must be powdered first) and only the cement in silicic rocks.

Other non-silicic rocks have fairly distinctive characteristics: **rock gypsum**, is very soft, **rock salt** tastes like halite, and **coal** is black and will burn (but please do not try this in class). Ironstone may, at times, be difficult to recognize because it can closely resemble hematite-cemented **sandstone** or **siltstone**.

Once the general composition has been recognized, you next need to determine whether the texture is **crystalline** or **clastic**. The former is the same texture found in phaneritic and aphanitic igneous rocks (the origins of the textures are, of course, very different: precipitation from water as opposed to crystallization from magma). Similarly, the clastic texture is identical to that of the fragmental igneous rocks. The terminology derives from the fact that another name for the grains in such rocks is "clasts." One rock type, **coal**, can have either a clastic texture (with plant fragments and other clasts) or an **amorphous** texture (that is, a homogeneous, non-crystalline mass lacking discernable clasts of any type).

The final distinctions among the sedimentary rock varieties are based on the specific aspects of texture and composition (that is, on the size, mineralogy, shape, and/or origin of the grains). A magnifying glass or binocular microscope may be needed when examining the finer-grained rocks. Note that the size class boundaries for clastic rocks in Part I of Table 3.1 have the following useful equivalencies: 2 mm is the thickness of a 5 cent coin, and 0.0625 mm is slightly less than the thickness of one of these pages (each page is actually about 0.1 mm thick). Note also that for **siltstone** and **shale**, the term **clay** applies to both a grain size (less than 0.004 mm) and a mineral. This somewhat confusing usage arose in geology because most clay mineral grains in sediments are smaller than 0.004 mm.

Exercise 3

Your instructor will provide you with a set of 15 sedimentary rock samples. Carefully examine each sample and <u>neatly</u> record its properties on a "Sedimentary Rock Data Sheet." Use the classification in Table 3.1 to determine the rock name. This table includes all the different rock types in your set (but the set does not necessarily include all the rock types in the table!).

Hand in the filled-out data sheets to your instructor. These will be graded and returned to you.

Sedimentary Rock Data Sheet

Student's Name _____

Sample Number	Texture	Composition	Other Properties	ROCK NAME

Chapter 4

Metamorphic Rocks

Note: *it is recommended that you bring a magnifying glass to class.*

Origin of Metamorphic Rocks

Any pre-existing igneous, sedimentary or even metamorphic rock (collectively referred to as **parent rocks**) will undergo significant changes in composition and/or texture when exposed to **high temperatures** and **high pressures**. This process is call **metamorphism** and normally takes place deep below the Earth's surface. There the pressure steadily increases with depth because of the increasing weight of the overlying rock.

Temperature also steadily increases and this phenomenon is known as the **geothermal gradient**. The Earth's core is heated to about 6000°C due to the radioactive decay of uranium and other elements. The heat escapes outwards, eventually reaching the Earth's crust where it is responsible for igneous activity and, at temperatures below the melting point of rocks, metamorphism. Such high metamorphosing temperatures can occur over broad regions of the crust.

High temperatures can also occur more locally near the contact with igneous intrusions. As these magma bodies cool they shed enormous amounts of heat into the surrounding rock. They also sometimes produce **hydrothermal fluids** (both water-rich magmas and super-hot waters enriched in dissolved minerals) that permeate the surrounding rock and cause other changes.

There are, thus, two broad categories of metamorphism: **regional**, and **contact** (see Figure 4.1). Contact-metamorphosed rocks show the effects of high temperatures plus also sometimes the effects of hydrothermal fluids.

Regionally metamorphosed rocks can be produced by either high temperatures or high pressures acting alone or together. The effects of pressure are responsible for the development of **foliation**, a distinctive type of structure found only in metamorphic rocks (see below for further discussion).

Metamorphic Rock Properties and Classification

Like the other rock groups, metamorphic rocks are classified on the basis of their texture and composition (see Table 4.1). In addition to these two properties, the presence or absence of foliation is also used in the classification. It is by this latter property that the most important distinction is made among the varieties of metamorphic rocks: **foliated** vs. **non-foliated**.

In foliated rocks, the constituent grains, which often have elongated or platy shapes, share a common orientation (that is, they are all aligned in the same direction). These rocks, consequently, tend to break parallel to this direction and so yield thin, flat slabs or chips. Depending on the form it takes (see description in Table 4.1), this breakage is called either **slaty cleavage** or **schistose cleavage**. In another type of foliation, referred to as **gneissic structure**, the different minerals tend to segregate themselves into separate layers. Such rocks have alternating, thin layers of light-colored minerals (mostly quartz and feldspar) and dark-colored minerals (mostly amphibole and biotite). They do not, however, show a pronounced tendency to break along these layers and so do not exhibit "rock cleavage" as in the other types of foliated rock.

If the metamorphosing temperature becomes too high then, of course, the parent rock will melt,

FIGURE 4.1 Types of Metamorphism

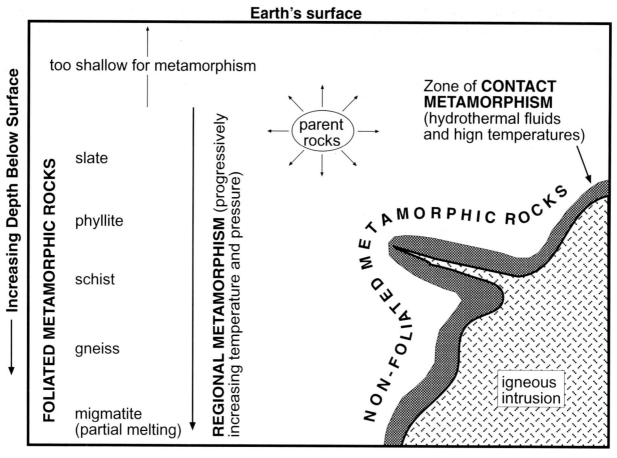

Earth's surface

Increasing Depth Below Surface

too shallow for metamorphism

FOLIATED METAMORPHIC ROCKS

slate

phyllite

schist

gneiss

migmatite
(partial melting)

REGIONAL METAMORPHISM (progressively increasing temperature and pressure)

parent
rocks

Zone of **CONTACT METAMORPHISM** (hydrothermal fluids and high temperatures)

METAMORPHIC ROCKS

NON-FOLIATED

igneous
intrusion

either entirely or partially. In the former case, a new magma is produced and we are back into the igneous realm, and in the latter case a hybrid rock, called **migmatite**, is created that is part metamorphic and part igneous.

Among the foliated rocks, there is a strong relationship between the type of foliation, and the **metamorphic grade** (that is, the temperatures and pressures of formation) (see Figure 4.1). The non-foliated metamorphic rocks are produced by contact metamorphism with some of these also forming at the lower to intermediate grades of regional metamorphism.

At the finer levels of classification, other textural and compositional characteristics are used to determine the rock name. The minerals in metamorphic rocks are mostly the same

"rock-forming" varieties you have encountered previously. Metamorphism can, however, produce some new minerals, like **garnet**, and you may see these in the rock samples.

TABLE 4.1					
Simplified Classification of Metamorphic Rocks					
Structure & Texture		Typical Characteristics	ROCK NAME	Probable Parent Rock	Type of Meta-morphism
Nonfoliated	psuedo-clastic	Sand-size grains: rock resembles sandstone but breaks through grains rather than around them. Shades of gray, brown or red.	**quartzite**	sandstone	contact or regional (low - intermediate grade)
		Pebble- or cobble-size grains: rock resembles conglomerate or breccia but breaks through grains rather than around them. Can be any color.	**meta-conglomerate or metabreccia**	conglomerate or breccia	
	amor-phous	Composed of organic carbon. Has conchoidal fracture. Black with a shiny/glassy surface.	**anthracite coal**	bituminous coal	regional (low - intermediate grade)
	phaneritic	Composed of calcite or, less often, dolomite. Mainly white, or shades of gray or pink.	**marble**	limestone or dolostone	contact or regional (low - intermediate grade)
	aphanitic	Composed of silicate minerals, mostly quartz and feldspar. Has dull luster. Mainly black, dark gray or green.	**hornfels**	siltstone, any volcanic igneous rock (esp. basalt)	contact
Foliated	crystalline	Slaty cleavage: rock splits into thin (<1 cm) flat slabs. Surface has dull to slightly glossy luster. Black, dark gray, green or red.	**slate**	shale	regional (low grade)
		Slaty cleavage: breaks into thin (<1 cm) flat to wavy slabs. Surface has glossy luster. Black, dark gray or green.	**phyllite**	shale or slate	regional (low - intermediate grade)
	phaneritic	Schistose cleavage: breaks unevenly along wavy foliations. Predominant minerals are micas: chlorite, muscovite or biotite. Garnet and other exotic minerals are commonly present. Silvery white, light to dark gray, or green.	**schist** (use predominant minerals in rock name; e.g., mica schist)	shale, slate or phyllite	regional (intermediate grade)
		Gneissic structure: alterna-ting layers of light- (quartz and feldspar) and dark-colored (horn-blende and biotite). Layers may be deformed. Garnet and other exotic minerals may be present.	**gneiss**	shale, plutonic igneous rocks (esp. granite), slate, phyllite, and schist	regional (high grade)
		Mixture of gneiss and plutonic igneous rock (usually granite). Has same colors as gneiss and granite.	**migmatite**	any rock	regional (very high grade) with partial melting

Exercise 4

Your instructor will provide you with a set of 8 metamorphic rock samples. Carefully examine each sample and <u>neatly</u> record its properties on a "Metamorphic Rock Data Sheet." Use the classification in Table 4.1 to determine the rock name. This table includes all the different rock types in your set (but the set does not necessarily include all the rock types in the table!).

Hand in the filled-out data sheets to your instructor. These will be graded and returned to you.

Metamorphic Rock Data Sheet

Student's Name _____

Sample Number	Structure	Texture	Mineralogy	Other Properties	ROCK NAME

Chapter 5

Geologic Time and Age Dating

Note: *you will be asked to do some simple calculations for one of the exercises in this chapter. Although an electronic calculator is not necessary, you may wish to bring one to class.*

Geologic Time

The Sun, planets and moons of our solar system formed through the "gravitational condensation" of interstellar debris. The process started sometime before 5 billion years (BY) ago and by about 4.6 BY ago the planet Earth had formed (that is, our planet had become a solid, rocky sphere).

Geologists are now attempting to identify and date all the major changes that have occurred in the Earth during the last 4.6 BY. They do this by studying rocks, which provide a record of these changes. From this **rock record** one can learn about the rise and fall of mountains, the expansion and contraction of oceans, the wanderings of continents, and the evolution and extinction of myriad varieties of plants and animals. How much can be learned about these and other facets of Earth history depends, of course, on the preservation of rocks from the different time periods.

Besides being able to "read" what rocks have to say, geologists also have to determine the age (time of formation) of rocks in order to understand Earth history. There are two very different but complementary approaches to dating rocks: **relative** and **absolute age dating**. The objective of relative age dating is to establish the <u>order</u> in which geologic events occurred in time. Absolute age dating, in contrast, tells us how many years ago the events occurred.

Relative age dating has been practiced by geologists for about 200 years but absolute age dating, which requires sophisticated analytical equipment, has been used only during the last half century. With the results of both taken together, geologists have created the combined **Geologic Column** and **Time Scale.** An abbreviated version of this chart is given in Table 5.1. Note that in this chart the different named segments of Earth history are represented by boxes, the heights of which are proportional to their relative durations (that is, they are drawn "to scale").

Relative Age Dating

Each layer or other rock body in the Earth's crust is a datable **geologic event**. For example, a sedimentary rock layer represents "deposition," and an igneous rock body represents either "intrusion" or "extrusion." Other geologic events commonly seen in rock sequences are "faulting" (displacement of rocks on opposite sides of a fracture), "deformation" (tilting or folding of rocks), and "erosion" (as indicated by unconformities; see below).

To establish the order in which events occurred, geologists apply several simple observational principles. These are illustrated in Figure 5.1. The **Principle of Original Horizontality** states that sedimentary rocks and volcanic igneous are originally deposited in horizontal (or nearly so) layers. If a rock layer is now tilted or folded, it must have experienced a post-depositional deformation event.

According to the **Principle of Superposition**, in a sequence of layered sedimentary or volcanic igneous rocks the overlying layers are younger than the underlying ones. In the case of plutonic igneous batholiths and stocks (but not dikes and sills), the overlying rock layers are assumed to be

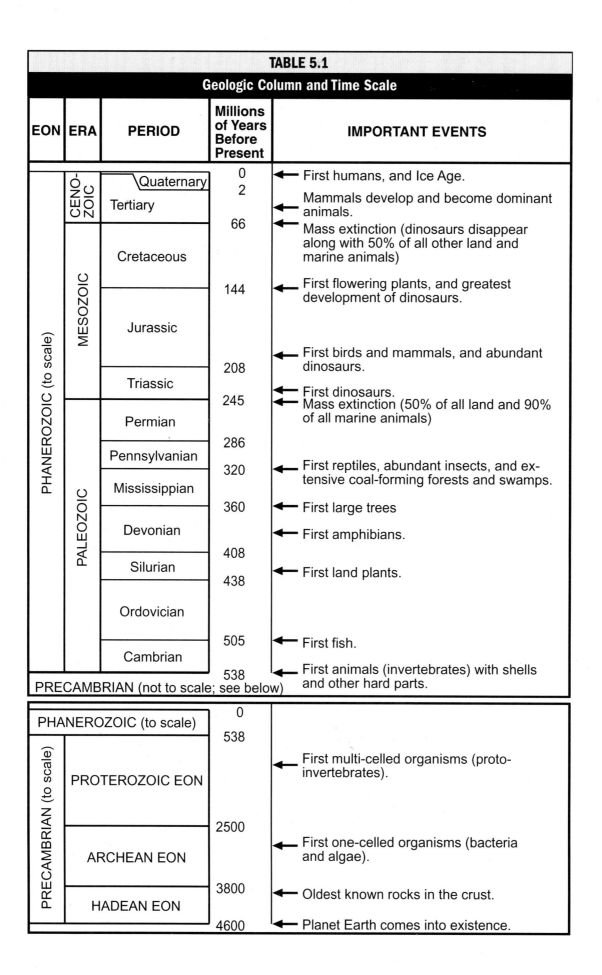

TABLE 5.1				
Geologic Column and Time Scale				
EON	**ERA**	**PERIOD**	**Millions of Years Before Present**	**IMPORTANT EVENTS**
PHANEROZOIC (to scale)	CENO-ZOIC	Quaternary	0	← First humans, and Ice Age.
		Tertiary	2	← Mammals develop and become dominant animals.
	MESOZOIC	Cretaceous	66	← Mass extinction (dinosaurs disappear along with 50% of all other land and marine animals)
			144	← First flowering plants, and greatest development of dinosaurs.
		Jurassic		
		Triassic	208	← First birds and mammals, and abundant dinosaurs.
	PALEOZOIC	Permian	245	← First dinosaurs. ← Mass extinction (50% of all land and 90% of all marine animals)
		Pennsylvanian	286	
		Mississippian	320	← First reptiles, abundant insects, and extensive coal-forming forests and swamps.
		Devonian	360	← First large trees
				← First amphibians.
		Silurian	408	← First land plants.
		Ordovician	438	
		Cambrian	505	← First fish.
			538	← First animals (invertebrates) with shells and other hard parts.

PRECAMBRIAN (not to scale; see below)

		Millions of Years	Important Events
PRECAMBRIAN (to scale)	PHANEROZOIC (to scale)	0	
		538	← First multi-celled organisms (proto-invertebrates).
	PROTEROZOIC EON		
		2500	← First one-celled organisms (bacteria and algae).
	ARCHEAN EON		
		3800	← Oldest known rocks in the crust.
	HADEAN EON		
		4600	← Planet Earth comes into existence.

FIGURE 5.1 Principles Used to Determine Relative Ages

A. PRINCIPLES of ORIGINAL HORIZONTALITY and SUPERPOSITION

Youngest E
     D
      C
       B
        A Oldest

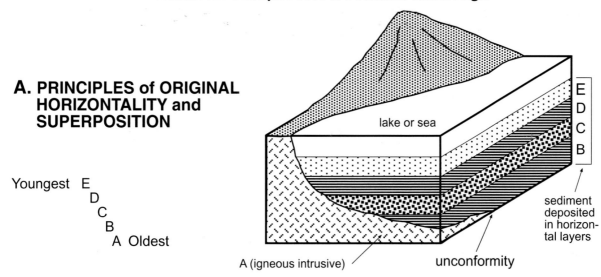

lake or sea

E D C B

sediment deposited in horizontal layers

A (igneous intrusive)

unconformity

B. PRINCIPLE of CROSS-CUTTING RELATIONSHIPS

Youngest J
     I
      H
       G
        F
         E
          D
           C
            B
             A Oldest

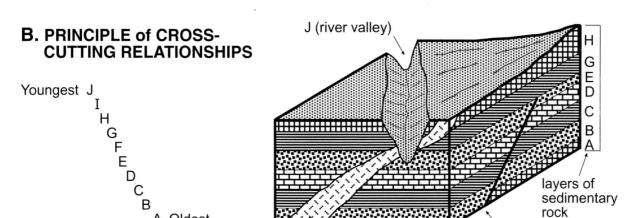

J (river valley)

H G E D C B A

layers of sedimentary rock

I (igneous intrusive)

F (fault)

C. PRINCIPLE of INCLUSIONS

Youngest J
     I
      H
       F, G
        E, G
         D
          C
           B
            A Oldest

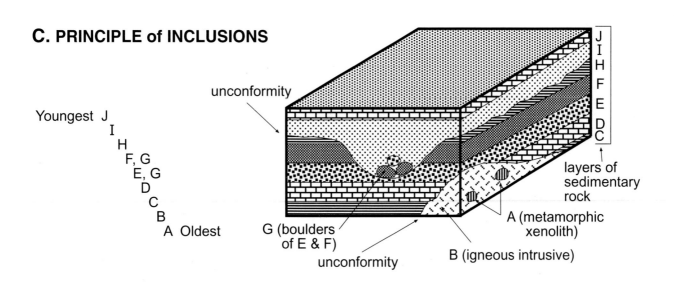

unconformity

J I H F E D C

layers of sedimentary rock

G (boulders of E & F)

unconformity

A (metamorphic xenolith)

B (igneous intrusive)

younger (as in parts A and C of Figure 5.1). This is usually, but not always, true in the real world. The question is easily resolved in the field by applying other criteria.

The **Principle of Cross-Cutting Relationships** says an igneous dike or sill, fault, or erosional surface is younger than the rocks it cuts. And finally, according to the **Principle of Inclusions**, rocks occurring as foreign inclusions (**xenoliths**) in igneous intrusives or as silicic clasts in sediments are older than the rocks in which they occur.

The above principles are the ones you will be applying in the exercises for this chapter. There are, however, others and the most important of these is the **Principle of Fossil Succession**: fossil plants and animals succeed one another in time in a definite and recognizable order. Thus, for example, a rock layer with dinosaur fossils is known to be older than one with human remains.

Unconformities are buried erosional surfaces and, as such, represent breaks or gaps in the geologic record. A **disconformity** separates sedimentary or volcanic igneous rock layers that are parallel to one another. It usually results from a geologically brief, local cessation of deposition either with or without accompanying erosion. Disconformities can also occur on a global scale when the level of the world ocean falls (with concomitant erosion of rocks/sediments stranded above sea level) and then rises (with renewed deposition on the drowned erosional surface).

An **angular unconformity** separates layered sedimentary or volcanic igneous rocks of different structural orientations. Most commonly there are flat-lying or horizontal layers above tilted or folded layers. Such major breaks in the rock record can be produced by regional tectonic uplift above sea level with concomitant deformation. The latter causes the tilting or folding, and the former causes the erosion.

A **nonconformity** separates sedimentary or volcanic igneous layers from underlying intrusive igneous or metamorphic rocks. It represents major regional tectonic uplift above sea level on a scale usually greater than that for angular unconformities (such as in "mountain building"). Only in major mountain-building events can uplift and erosion expose at the surface the plutonic igneous and metamorphic rocks that make up the deeper parts of the continental crust.

Tectonic uplift creates the opportunity for erosion in angular unconformities and nonconformities, but to get deposition on the erosional surface another geologic event must occur. This could be either tectonic subsidence, a change in climate, or a rise in global sea level.

FIGURE 5.2 Types of Unconformities

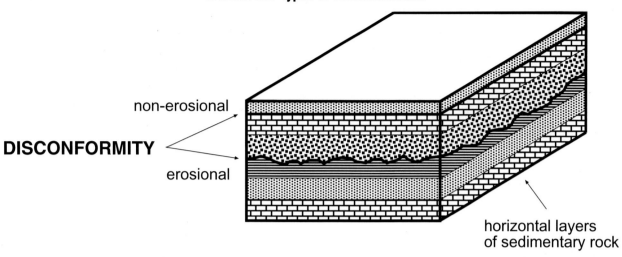

DISCONFORMITY

non-erosional

erosional

horizontal layers
of sedimentary rock

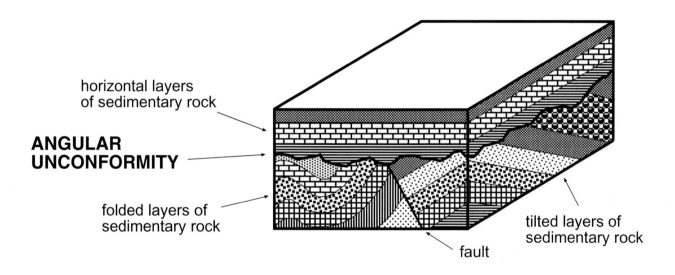

horizontal layers
of sedimentary rock

**ANGULAR
UNCONFORMITY**

folded layers of
sedimentary rock

tilted layers of
sedimentary rock

fault

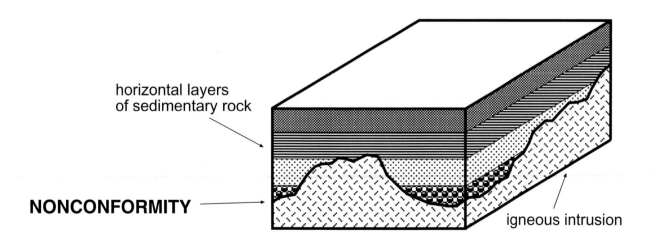

horizontal layers
of sedimentary rock

NONCONFORMITY

igneous intrusion

Exercise 5A

Figures 5.3 and 5.4 provide hypothetical cross-sections of portions of the Earth's crust. Each consists of numerous rock bodies and other datable events. Your task is to determine the _precise_ order in which these events occurred. Using the "Answer Sheet for Relative Age Dating," list the lettered events for each cross-section in order of increasing age downward. Start with Figure 5.3A, which has the easier cross-sections.

Hand in the filled-out answer sheet to your instructor. This will be graded and returned to you.

FIGURE 5.3 Exercises in Relative Age Dating (Part 1)

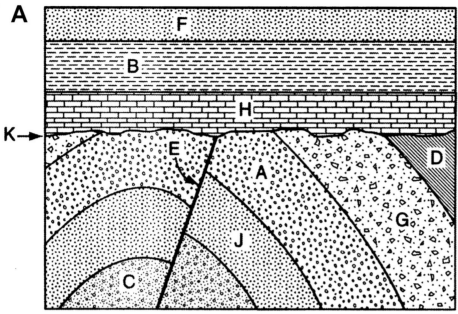

Note: K is an unconformity, E is a fault, and all rock layers are sedimentary or volcanic igneous.

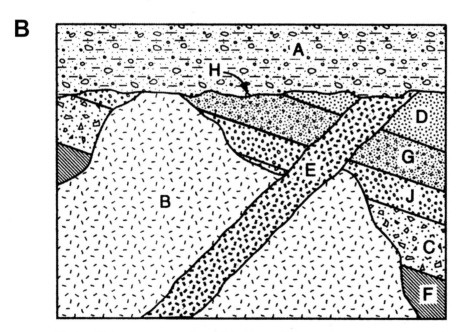

Note: H is an unconformity, B and E are igneous intrusions, and all other rock layers are sedimentary or volcanic igneous.

Figures from Physical Geology Exercises, 2nd edition by Jennifer Hinds. Copyright © 2004 by Kendall Hunt Publishing Company. Reprinted by permission.

FIGURE 5.4 Exercises in Relative Age Dating (Part 2)

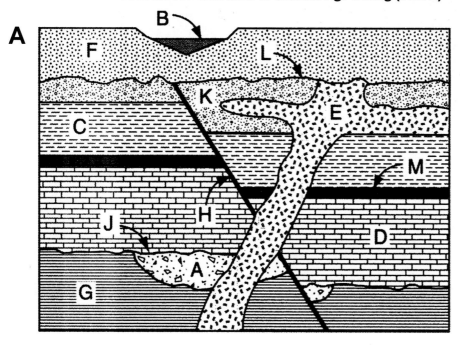

A

Note: J and L are unconformities, H is a fault, E is an igneous intrusion, and all rock layers are sedimentary or volcanic igneous.

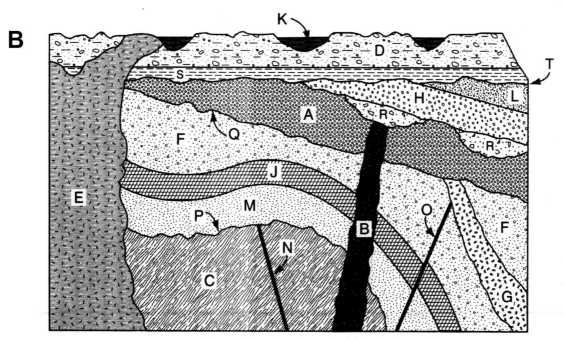

B

Note: P, Q and T are unconformities; B, E and G are igneous intrusions; N and O are faults; C is metamorphic; and all other rock layers are sedimentary or volcanic igneous.

Absolute Age Dating

Absolute age dating is also known as **radiometric dating** because it uses the radioactive elements in rocks. You will recall that all atoms having the same atomic number (that is, the same number of protons in the nucleus) are the same element. The number of neutrons in the nucleus of a given element can, however, vary slightly from atom to atom. Each subvariety of an element, created by the different numbers of neutrons, is called an **isotope** of that element. Each elemental isotope is uniquely identified by its **atomic mass**, which is equal to the total number of protons plus neutrons in the nucleus. Consider, for example, the element potassium (with symbol K). It has 19 protons but can have either 20 or 21 neutrons. There are, thus, two isotopes of potassium and these are represented symbolically as K-39 and K-40. The numbers are the atomic masses.

The isotopes of some elements are inherently **unstable** and so spontaneously decay to a different element. They do this by ejecting pieces of the nucleus, which has the effect of changing the atomic number. The ejected pieces are called **radiation** and the decaying atoms are said to be **radioactive**. The original radioactive isotope is known as the **parent** atom and each new element produced by its decay is a **daughter** atom. A given radioactive isotope may decay to a single daughter or to a whole series of daughters (each of which is radioactive itself and so decays to the next daughter in the series). The decay process stops when the last daughter atom produced is **stable** (that is, not radioactive).

The rate at which a given radioactive isotope decays is constant over time and is referred to as its **half-life**. The half-life concept is illustrated in Figure 5.5. Let us use the decay of uranium-238 (or U-238) as an example. This isotope has a half-life of 4.5 billion years, which is only by a coincidence approximately equal to the age of the Earth. What this means is that if we start with 100 U-238 atoms then after 4.5 BY there will only be 50 left. The other 50 atoms are now various daughter isotopes. After another 4.5 BY (now 9 BY since the beginning) there will only be 25 U-238 atoms left

(half the number remaining after the first 4.5 BY). Decay continues with the remaining U-238 atoms converting to daughters at the same half-life rate as shown in Figure 5.5.

Half-lives are easily determined in the laboratory, and are known to be unaffected by the temperatures, pressures and chemical environments to which rocks are subjected in the Earth. They are, thus, truly and absolutely "constant" and because of this radioactive isotopes can be used for age dating. They are, in short, what makes the "radiometric clocks" tick.

One of the most useful clocks is that based on U-238. It can be used, for example, to date the time of crystallization for a granite pluton. Reading the clock involves the following steps. First, a single grain of zircon is collected from the rock. This is a scarce but nearly ubiquitous mineral in granites. Second, an instrument called a "mass spectrometer" is used to count the number of U-238 atoms in the zircon grain as well as the number of daughter atoms produced by U-238's decay. Third, knowing U-238's half-life and the relative amounts of the parent (U-238) and daughter atoms, one can calculate the age of the zircon grain and, hence, the granite. This works because the U-238 clock starts "ticking" as soon as the zircon grain crystallizes from the magma, trapping within it a quantity of U-238 atoms which are always present as an impurity.

The actual calculation is fairly involved but it can be simplified, for approximate ages, to the following formula:

$$\text{age (in years)} = 1.443 \cdot T_{1/2} \cdot (N_d/N_p)$$

where 1.443 is a constant, $T_{1/2}$ is the half-life (in years), N_d is the number of daughter atoms present, and N_p is the number of parent atoms present. Consider the following worked example using the U-238 clock. Suppose a zircon grain has 27,300 parent (U-238) atoms and 3,800 daughter atoms. The ratio N_d/N_p equals 0.139, and this number multiplied by 1.443 and then by 4.5 BY yields an age of 903 million years. This then is the age of the granite containing the zircon. The number of half-lives that have elapsed since this crystallization event is 0.201 (that is, 0.903 BY divided by 4.5 BY).

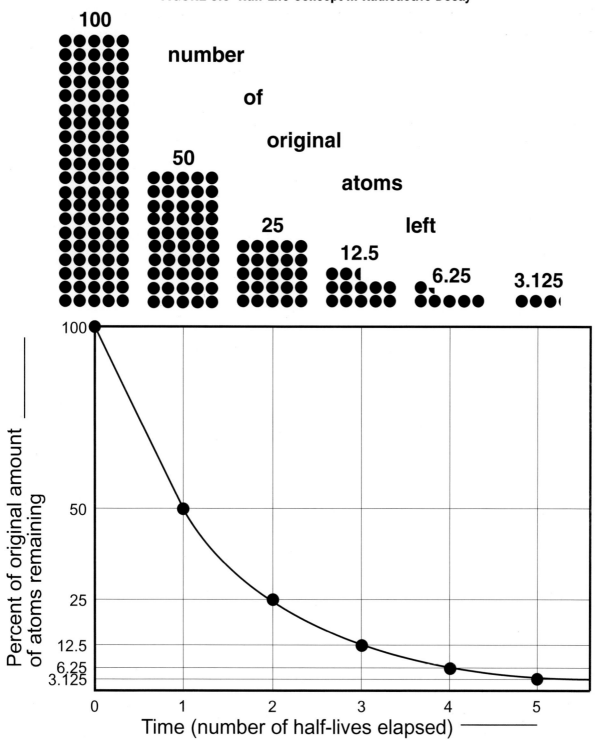

FIGURE 5.5 Half-Life Concept in Radioactive Decay

In actual practice, numerous zircon grains from different parts of a granite pluton would be individually dated and the results averaged. It is also common to verify dates by using more than one type of radiometric clock. Thus, a granite may be dated by both the U-238 and K-40 methods.

The big drawback to radiometric dating is that it can only be used on igneous rocks (for the time of crystallization) and some metamorphic rocks (for the time of recrystallization). The absolute age of sedimentary rocks cannot be measured. However, if there are datable igneous/metamorphic rocks both older and younger than a sequence of sedimentary layers (as established by relative age dating), then the absolute age of the latter is known to fall between the two absolute ages of the former.

Exercise 5B

Several questions are asked below and for these you need to write your responses on the "Answer Sheet for Absolute Age Dating." All of the questions relate to cross-section B in Figure 5.4.

1. In igneous intrusion "G", a granite dike, three grains of zircon have been analyzed using the U-238 clock. The numbers of parent (N_p) and daughter (N_d) atoms are given on the answer sheet. Compute and fill in the intermediate calculation results and age for each grain, and then compute the average age. Be sure to use U-238's half-life of 4.5 BY.

2. In igneous intrusion "B", a diorite dike, three grains of hornblende have been analyzed using the K-40 clock. Do all of the same calculations that you did for the first question and record your responses on the answer sheet. This time, however, you must use K-40's half-life, which is 1.3 BY.

3. How many half-lives have elapsed since intrusions "B" and "G" crystallized? (use their average ages)

4. During what geologic periods were intrusions "B" and "G" emplaced? (see Table 5.1)

5. During what geologic period(s) could sedimentary layers "A", "F" and "R" have been deposited?

6. What geologic period represents the oldest possible age for sedimentary layer "L."

    Carry out all calculations to at least the fourth digit to the right of the decimal point.

    Hand in the filled-out answer sheet to your instructor. This will be graded and returned to you.

Answer Sheet for Relative Age Dating

Exercise A in Figure 5.3 (A–K, no I)

Youngest _____

_____

_____

_____

_____

_____

_____

_____

_____

Oldest _____

Exercise A in Figure 5.4 (A–M, no I)

Youngest _____

_____

_____

_____

_____

_____

_____

_____

_____

_____

_____

Oldest _____

Exercise B in Figure 5.3 (A–J, no I)

Youngest _____

_____

_____

_____

_____

_____

_____

_____

Oldest _____

Exercise B in Figure 5.4 (A–R, no I)

Youngest _____

_____

_____

_____

_____

_____

_____

_____

_____

_____

_____

_____

_____

_____

_____

_____

Oldest _____

Answer Sheet for Absolute Age Dating

1.

Grain	N_p	N_d	N_d/N_p	$N_d/N_p \bullet 1.443$	$N_d/N_p \bullet 1.443 \bullet T_{1/2}$	AGE
1	123,898	8,519	_____	_____	_____	_____
2	87,100	6,304	_____	_____	_____	_____
3	111,275	8,457	_____	_____	_____	_____
					Average	_____

2.

Grain	N_p	N_d	N_d/N_p	$N_d/N_p \bullet 1.443$	$N_d/N_p \bullet 1.443 \bullet T_{1/2}$	AGE
1	201,512	10,715	_____	_____	_____	_____
2	134,509	7,529	_____	_____	_____	_____
3	177,336	10,422	_____	_____	_____	_____
					Average	_____

3. "G" _____   "B" _____

4. "G" _____   "B" _____

5. "A" _____   "F" _____

   "R" _____

6. _____

Topographic Contour Maps

Note: *several simple calculations are required for the exercises in this chapter and so you may want to bring an electronic calculator to class. You may also want to bring a long metric ruler. A ruler of sorts is provided in Table 6.1 but a separate ruler would be more useful.*

Introduction

The Earth's surface is not flat and featureless, but rather exhibits a rich variety of landscapes. These are collectively referred to as the Earth's **topography**. Geologists (and geographers) make **topographic maps** to show the size, shape, and distribution of landscape features. These maps are necessarily "planimetric" (that is, two-dimensional) but must somehow show the three-dimensional configuration of the surface. There are several ways of doing this, but the most useful and accurate is with **contour maps**.

About Contour Maps

An **elevation contour** is a line on a map along which every point has precisely the same elevation (that is, the same vertical distance above the average global sea level). Sea level, by definition, has an elevation of 0 feet. An example of a contour line is the "water's edge" around Ohio's Lake Erie. The lake's level rises and falls with the years, seasons, and atmospheric conditions. However, at any given instant on a windless day, the water's surface is effectively horizontal and, thus, intersects the land surface everywhere at the same elevation (about 571 feet above sea level).

Maps can use contour lines to show topography. The lines always have fixed elevations where the **contour interval** (that is, the elevation difference between adjacent lines) is constant. An example of

such a map is shown in Figure 6.1. The perspective drawing represents the topography seen in the underlying map. The area is on a sea coast and so the water's edge (shoreline) is the 0 foot contour line. The higher elevations are depicted by contours having a 20 feet interval (20, 40, 60, etc., up to 280 in the upper right corner). For easier visualization, it might help to think of the contours as marking the successive shorelines if the ocean were to rise in increments of 20 feet.

It is conventional for published contour maps to have every 5th contour drawn as a thicker line (as seen in Figure 6.1) and normally only these have the elevation printed on them. If the contour interval is constant (as it always should be), it is easy to determine the elevations of the unlabeled contours. Thus, for example, we know that the rectangular structure on the right side of the map in Figure 6.1 sits astride the 60 feet contour line.

An alternative to contour lines on maps is to use a different color for each elevation interval. Such depictions of topography are basically just "colorized contour maps." They are fine for illustrating the general aspects of topography but are impractical for showing the topographic details.

Figure 6.1 shows another important aspect of topographic contour maps: they also depict "water features" (streams and rivers, ponds and lakes, swamps, and ocean shorelines), and "cultural features" (railroads and highways, towns and cities, sometimes individual buildings and other man-made structures, and land boundaries). They also commonly show forests and croplands.

Topographic contour maps are now available for every part of the world! For areas in the United States they are produced and sold by the U. S. Geological Survey (USGS), and for areas elsewhere in the world they can be obtained from the U. S.

TABLE 6.1

Units of Measure in the English and Matric Systems, and Conversions From One to the Other

ENGLISH UNITS OF LINEAR MEASURE

12 inches = 1 foot

1 mile = 5280 feet = 63,360 inches

Abbreviations: in = inches, ft = feet, and mi = miles

METRIC UNITS OF LINEAR MEASURE

10 millimeters = 1 centimeter

100 centimeters = 1 meter

1000 meters = 1 kilometer

Abbreviations: mm = millimeters, cm = centimeters, m = meters, and km = kilometers

CONVERSION OF ENGLISH UNITS TO METRIC UNITS

when you know	multiply by	to find
inches	25.4	millimeters
inches	2.54	centimeters
feet	30.48	centimeters
feet	0.3048	meters
miles	1.609	kilometers

CONVERSION OF METRIC UNITS TO ENGLISH UNITS

when you know	multiply by	to find
millimeters	0.03937	inches
centimeters	0.3937	inches
meters	3.2808	feet
kilometers	0.6214	miles

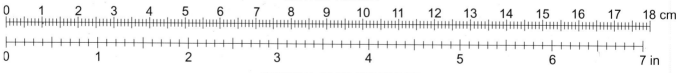

METRIC RULER

DECIMAL ENGLISH RULER

FIGURE 6.1 Example of Topographic Contour Map

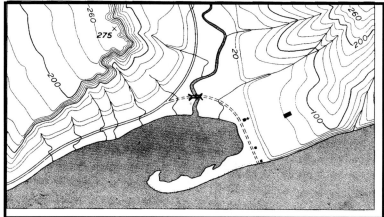

Courtesy of the U. S. Geological Survey

Defense Mapping Agency (USDMA). The Map Department on the third floor of Carlson Library has virtually all the maps (thousands of them) published by the USGS and USDMA as well as a great many more from other sources. You might enjoy visiting the map collection and if you see some maps that you would like to have, the Library staff can tell you how to order them.

When topographic contour maps were first made by the USGS in 1879, he work was done by surveyors in the field. Beginning in the late 1940's, however, the maps were made from photographs taken from airplanes (**aerial photographs**; see Chapter 7). Aerial photographs of the same area taken from slightly different vantage points can be viewed together to produce a three-dimensional (**stereoscopic**) image that, with the aid of sophisticated instrumentation and some known elevations on the ground, can be converted to a contour map. Such maps are also now being made from satellite photographs and all of those from the USDMA are of this type.

Topographic contour maps are an essential tool for geologists who use them to study landforms and to construct geologic maps (that is, maps showing rock units and the deformation structures within them). Most of the maps, however, are used by non-geologists for a wide variety of applications, including engineers planning construction projects, agricultural scientists doing soil surveys and crop estimates, military strategists planning troop movements and dispositions, airplane pilots looking for navigational landmarks, environmental scientists surveying the extent of pollution, hikers and hunters locating themselves in the wilderness, and many others.

Reading topographic contour maps takes practice, and the best way to learn is to make some. This you will do using the simple methods of the old-time surveyors.

Making a Contour Map

The process of making a contour map begins by determining the precise elevations of numerous points within an area of interest (as seen in Figure 6.2). These are referred to as **spot elevations**.

FIGURE 6.2 Example of Spot Elevations used in Contour Mapping

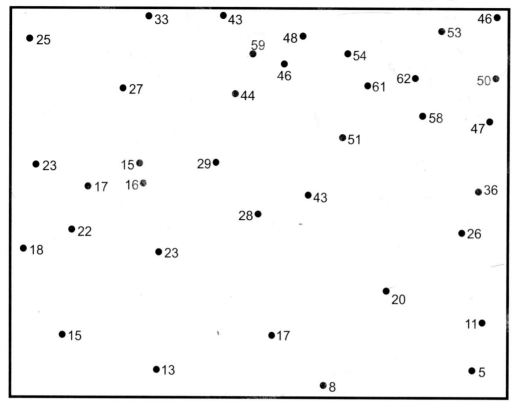

Next you have to decide on a contour interval, and a starting value for the first or lowest contour line. The interval selected will depend mainly on the total elevation range in the area. This quantity is known as the **relief** and is computed by subtracting the elevation of the lowest point from that of the highest point. If an area has, say, 500 feet of relief you would not want to use a contour interval of 5 or even 10 feet because this would result in 100 or 50 different contours, respectively. These are too many and would produce a very crowded-looking map (not to mention one that would take a very long time to draw). You also have to guard against having too few contours because then the topographic details are lost. The optimal contour interval ultimately depends not only on the relief but also on the number of spot elevations available: the more elevations there are, the more topographic details can be shown, and so the smaller the contour interval should be. The intervals used on different USGS maps are commonly 10, 20, 50 or 100 feet (always, you will notice, a multiple of 10 feet). Most USGS maps use

"feet" but map makers ("cartographers") just about everywhere else in the world use "meters."

The value for the first (lowest) contour is set as close as possible to the lowest spot elevation in the map area but, at the same time, must be a number that when divided by the contour interval yields an integer value (that is, a whole number). If, for example, an area has a minimum spot elevation of 87 feet, an appropriate starting contour would be 90 for a contour interval of 10 feet, or 100 for contour intervals of 20 or 50 feet. For the exercises in this chapter, the contour intervals and starting contour elevations will be given to you.

The next and most important step in making a map is drawing the contour lines. One begins by locating the highest and lowest points, and trying to get a "sense" of what the topography is doing in between. You can then start drawing the contours. In doing this there are numerous rules that must be followed, and these are described below and illustrated in Figures 6.3 and 6.4.

1. Contour lines must "honor" all the spot elevations. This means that each of these

FIGURE 6.3 Contoured Spot Elevations from Figure 6.2

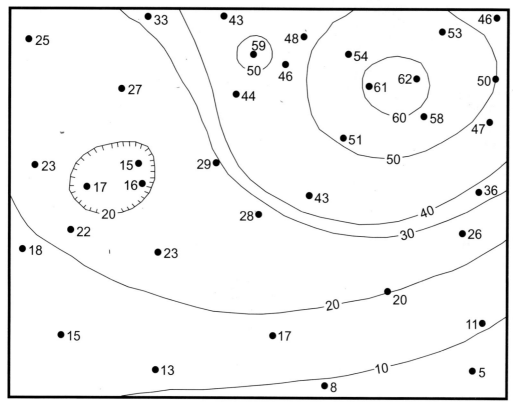

Contour Interval = 10 feet

FIGURE 6.4 Map Illustrating Some of the Rules of Contouring

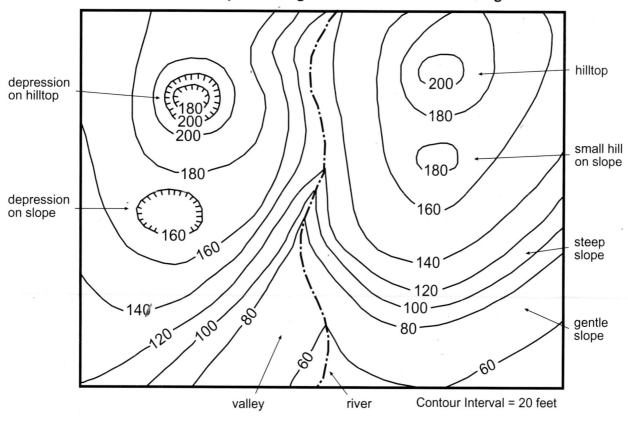

Contour Interval = 20 feet

elevations must either fall between two contour lines having higher and lower elevations, or on a contour line if it has the same elevation.

2. Contour lines should, in general, pass closer to spot elevations with similar values than to those with less similar values. For example, a 50 feet contour line should come closer to a spot elevation of 48 feet than to one of 43 feet, all else being equal. You, thus, need to practice a kind of "visual interpolation" when drawing contours between spot elevations.

3. All contour lines between the lowest and highest elevations must be used on a map (that is, you cannot omit certain contours).

4. Contour lines in the "real world" always close to form an irregular loop (thus, if you walked along one you would eventually get back to where you started). This "closure," however, will not always be visible or possible in the map area because the contours are artificially truncated at the map boundaries.

5. Two contour lines cannot cross each other, and a single contour line cannot split into two lines. For lines to do such things would require the impossible situation where the point at which lines cross or split has two different elevations simultaneously.

6. Contour lines can merge into a single line but only where they are representing a vertical cliff. In such a case, the lines are more appropriately thought of as being stacked one on top of the other.

7. Contour lines will be more closely spaced where the ground surface has a steeper slope and more widely spaced where slopes are gentler. Evenly spaced contours indicate a uniform or constant slope.

8. Where contour lines cross a stream valley they always form a **V** pointing upstream with the apex of the **V** on the stream itself. That they should do this is usually not obvious from the spot elevations, but if contours were actually drawn on the ground they would always appear as just described.

9. A concentric series of closed contours represents either a hill or a depression. A depression, of course, is a low area that is completely surrounded by higher ground. In order to distinguish it from a hill, the contour lines that fall <u>inside</u> the depression have hachure marks pointing inward.

10. A small hill on sloping ground will have its lowest contour repeated immediately upslope.

11. A depression on sloping ground will have its highest hachured contour repeated immediately downslope. Similarly, when a depression occurs at the top of a hill (as in the case of a volcanic crater) its highest hachured contour is repeated (but unhachured) on the immediately surrounding hilltop.

12. Contour lines should be smoothed out when drawn to make them look more natural, and they should also be kept reasonably parallel to each other where possible. In addition, contour lines should always be labeled with their elevations.

Even if all the above rules are followed, no two people will make identical contour maps. Differences in map "interpretation" will, however, decrease as the number of spot elevations increase.

Just for Fun

A **stereographic contour map** is one which, when viewed through special glasses, appears to be three-dimensional. It is as if you are looking down from an airplane onto a landscape where the contour lines are painted on the ground! Such maps are a useful and entertaining way to see the relationship between elevation contours and the three-dimensional topography they represent. You will notice that where there are depressions shown on these maps the contour lines are not hachured — they do not need to be if you can see depth!

Look at the two stereographic contour maps on the classroom wall. One shows the Grand Canyon in Arizona, and the other shows the Mount St. Helens volcano in Washington. These are a type of image known as an **anaglyph**. They are created by overlaying with a slight offset two contour maps, each of which has a slightly different vertical perspective. One map is printed in blue and the other in red. Look at the maps through the colored glasses on a wire attached to the wall. The blue lens goes over your right eye and the red one goes over your left eye (you can keep your regular glasses on). When looking through the colored glasses your right eye sees only the red contour map and your left eye sees only the blue one. Your brain integrates the two images and tricks you into thinking you are seeing a three-dimensional surface.

Also on one of the classroom walls there is a colorized contour map of Lake Erie. Here a gradational color scale is used to show different elevation intervals. The colors are specifically selected, however, to produce a 3-D effect: the warmer colors (reds, oranges and yellows) appear to be higher than the cooler colors (greens and blues). Many people will see this so-called "chroma-depth" effect without special glasses but for those who cannot, some glasses are provided (on a wire attached to the wall). The micro-optics incorporated into the glasses create two stereoscopic images from a single one. They do this by shifting the image colors in different directions for each eye.

For all these maps there is considerable "vertical exaggeration." This means that the apparent relief is greater than what you would see if looking at the same area from an airplane. More will be said about stereo viewing in Chapter 8 when you will be looking at stereo aerial photographs.

Making a Topographic Profile

It is sometimes useful to illustrate the topography of an area as a vertical profile along a straight line crossing the map. Such a **topographic profile** shows what you would see if it were possible to cut vertically down through the Earth's crust and then view the result from the side as a skyline silhouette. Figure 6.5 illustrates how a profile is made for the line across the contour map between A and B.

FIGURE 6.5 Example of a Topographic Profile

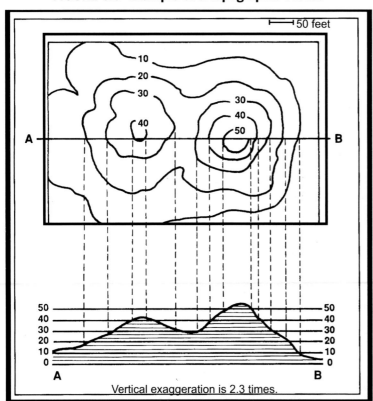

Courtesy of the U. S. Geological Survey

One starts by creating a graph where the equally spaced horizontal lines represent the contours on the map. You can, of course, use a vertical scale that is equal to the horizontal one on the map, but a certain amount of vertical exaggeration is desirable in order to emphazie the variations in topography along the profile. Wherever a contour crosses line A-B, draw a vertical (perpendicular) line down to the corresponding horizontal line (the one with the same elevation value). After doing this for all the contours crossing line A-B you will have a series of points. Connect these points with a smooth line and allow for elevations that are higher and lower than the contours on the hilltops and valley bottoms, respectively.

The amount of vertical exag-geration in the profile is determined by dividing the vertical distance (measured, preferably, with a metric ruler) between the upper and lower contour lines in the graph by the horizontal distance (again, measured with a ruler) on the map corresponding to the same number of feet as in the measured elevation difference on the graph. In Figure 6.5, for example, the distance between the 0 and 50 feet contours on the graph is 16 millimeters, and 50 feet on the map (see the upper right corner) is 7 millimeters. Dividing the first distance by the second (16/7) gives a vertical exaggeration of 2.3 times.

If you do not have a ruler, you can make use of the graphic scale at the bottom of Table 6.1. To use it, mark off a distance along the edge of a sheet of paper, and then place the edge on the scale and read off the distance.

Calculating Slope Gradient

Perfectly horizontal ground is a rarity. The Earth's surface almost always has some degree of inclination or slope, and this is known as its **gradient**. The gradient between two points is the ratio of the elevation difference (measured in feet or meters) divided by the distance between them as measured on a map (in miles or kilometers). For example, if two localities A and B are separated by 373 feet in elevation and 26.5 miles in distance, then the average slope of the ground between A and B has a gradient of 10.2 feet per mile (i.e., 373/36.5). Gradients can also, of course, be calculated in 'meters per kilometer.' Whatever the units employed, the gradient is an expression of the 'rate of descent' (or ascent) of the ground.

It is for rivers (and streams) that gradients are most often calculated. Note, however, that the distance between two points on a river is the actual length of the sinuous river channel, rather than a simple straight-line distance, between the two points.

Exercise 6A

1. Below, match the small contour maps on the right with the correct topographic profile on the left (write the number of the map beside the corresponding profile).

2. Contour Map 1 using a contour interval of 20 feet and a first contour line at 20 feet. Label all the contour lines. The dashed lines are streams.

3. On Map 2 write in the elevations where you see gaps in the contour lines. There are 14 such gaps. Next, draw a topographic profile for line X-X'. A ruler would be helpful for dropping the points vertically from the map to the profile but you can also use the straight edge on a sheet of paper. Write in the contour elevations (in meters) on the left side of the profile. Fill in the answers to the following questions below the profile: (i) what is the vertical exaggeration in the profile (measure the vertical and horizontal distances to the nearest half millimeter and compute to the nearest fraction); (ii) what are the elevations at points A and E; and (iii) what is the distance between points B and D?

MATCHING FOR EXERCISE 6A

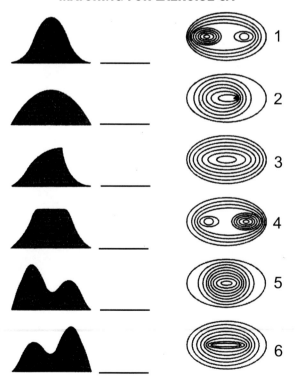

MAP 1 FOR EXERCISE 6A

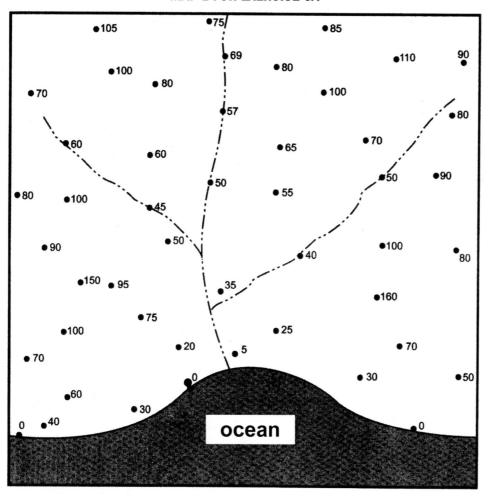

Figures from Physical Geology Exercises, 2nd edition by Jennifer Hinds. Copyright © 2004 by Kendall Hunt Publishing Company. Reprinted by permission.

MAP 2 FOR EXERCISE 6A

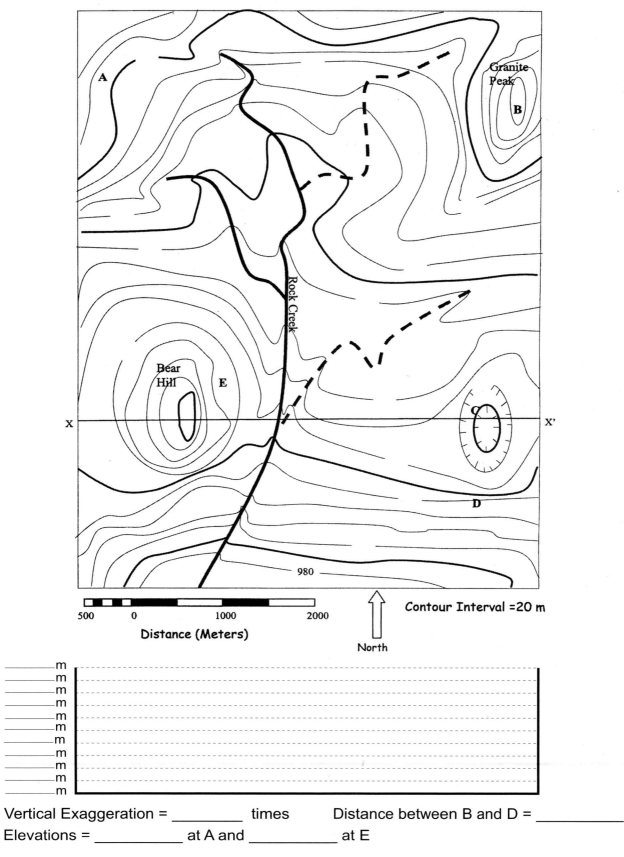

Contour Interval =20 m

Distance (Meters)

North

Vertical Exaggeration = _____ times   Distance between B and D = _____

Elevations = _____ at A and _____ at E

From laboratory Manual for Physical Geology by E. Baer et al, 1999.

13. Contour Map 3 using a contour interval of 20 feet. The first contour line is already drawn at 110 feet and the other solid line is a stream. Label all the contour lines.

Hand in the three maps and profile to your instructor. These will be graded and returned to you.

MAP 3 FOR EXERCISE 6A

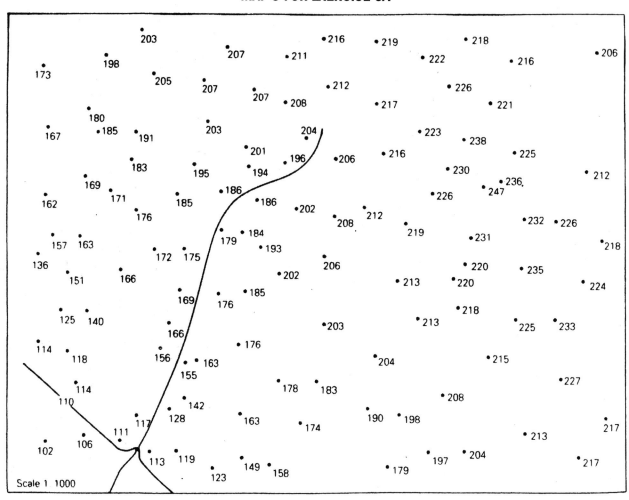

From Geology in the Laboratory by N. Gioppo, 2004.

Working with Real Contour Maps

For the last exercise in this chapter you will use the topographic contour maps at the back of this manual. Only small portions of larger maps published by the U. S. Geological Survey (USGS) are included here and much of the information printed around the margins of the complete maps is, unfortunately, omitted. Also in the back of this manual is a color page illustrating the topographic map symbols used by the USGS.

An example of a complete topographic map can be seen on one of the classroom walls. This map shows the Chief Mountain area in Montana. The work that went into making it is impressive when you realize that it was made entirely from field surveys (no aerial photographs were used). Beside the paper map is a plastic one of the same area, but this time the topography is also shown in "raised relief." Such raised-relief maps are commonly used as an educational tool but are impractical for use in the field.

Although the maps in this manual are incomplete, they do provide the essential information and this is printed below each map: graphic and proportional scales, contour interval, latitude and longitude for the map edges, north arrow, and name of the USGS map.

Scale

The "scale" of a map is given in two ways. Firstly, there are the **graphic (or bar) scales** for miles and kilometers. Such ruler-like scales are found on all maps and make it possible to directly measure horizontal distances between points. Secondly, a **proportional (or ratio) scale** is also provided. One used for some of the maps in this manual is "1:24,000." This notation means that 1 unit of distance (either English or metric) on the map equals 24,000 of the same units on the ground in the real world. Thus, a line 1 inch long on the map represents a distance on the ground of 24,000 inches (or 2000 feet or 0.379 miles; see Table 6.1 for conversions of units). The graphic and proportional scales are, of course, compatible with each other.

One can also express the above information as a **verbal scale**. For example, a given map could be said to have a scale of "1 inch equals 1 mile" or, equivalently, "one inch to the mile." This means that one inch on the map represents one mile on the ground. This is easily translated into a proportional scale: 1 mile equals 63,360 inches (see Table 6.1) and, thus, the proportional scale is 1:63,360. Converting from a proportional to a verbal scale is also easily done. For example, given a fractional scale of 1:125,000, the corresponding verbal scale in metric units is "1 centimeter equals 1.25 kilometers" (because 125,000 cm = 1,250 m = 1.25 km; see Table 6.1).

Maps can be described in relative terms as being either "small scale" or "large scale." For example, one at 1:24,000 is a larger scale map than one at 1:125,000. The terminology seems backwards to the numbers, but what it really means is that a one mile graphic scale is physically larger on a 1:24,000 map than it is on a 1:125,000 map.

Latitude and Longitude

The Earth's surface can be divided into a series of imaginary lines called **meridians** and **parallels** (see Figure 6.6). Meridians are circles drawn around the globe that pass through both the north and south geographic poles. It is through these poles that the Earth's rotational axis passes. Meridians are numbered according to their position relative to the "prime or zero meridian," which by international agreement passes through the Greenwich Observatory near London, England (this is also called the "Greenwich meridian"). As you may recall from high school geometry, a circle can be divided into 360 **degrees**, and each degree can be divided into 60 **minutes**, and each minute can be divided into 60 **seconds**. The "angular distance" of a meridian west or east of the prime meridian is, thus, given in degrees, minutes and seconds, and is referred to as the meridian's **longitude** (see Figure 6.6-C).

Parallels are also circles and, as their name suggests, are parallel to each other, unlike the meridians. Also unlike meridians, they do not cross each other and are perpendicular to the Earth's rotational axis. They are numbered according to

FIGURE 6.6 Latitude-Longitude Coordinate System Based on Meridians and Parallels

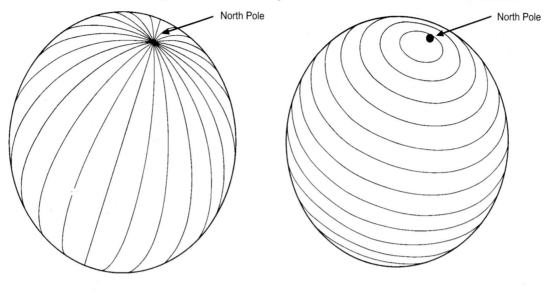

A. Meridians

B. Parallels

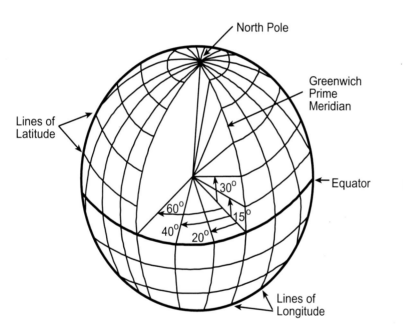

C. Angular Distances for Latitude and Longitude

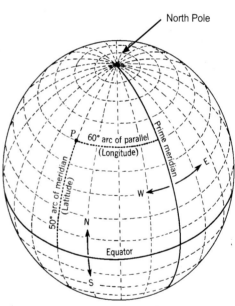

D. Determination of Latitude and Longitude for Point P

From *Introduction to Physical Geography* by Arthur N. Strahler, 1965.

Physical Geology Laboratory Exercises (EEES 1020)

their position relative to the Earth's equator. The angular distance of a parallel north or south of the equator is given in degrees, minutes and seconds, and is referred to as the parallel's **latitude** (see Figure 6.6-C).

Every point on the Earth's surface is crossed by a meridian and a parallel. These lines, thus, constitute a coordinate system that can be used to locate points in terms of latitude and longitude. In Figure 6.6-D, for example, point "P" has a longitude of exactly 60 degrees, and a latitude of exactly 50 degrees. Note that the longitude is measured along a parallel, and the latitude is measured along a meridian. Writing out the angular value of a latitude or a longitude can produce a wordy result unless some symbolic abbreviations are used. For example, if the value is 70 degrees, 14 minutes and 37 seconds, this can be written more compactly as 70°14'37". Latitudes in the U. S. are always north of the equator and longitudes are always west of the prime meridian. Thus, the latitude and longitude for Bowman-Oddy Laboratory on the University of Toledo campus would be written, respectively, as 41°39'36" N and 83°37'4" W.

For the maps in this manual, the approximate latitudes given are for the top/north and bottom/south edges of maps, and the approximate longitudes are given for the left/west and right/east edges. The left and right sides of the maps, thus, have an approximately north-south orientation. The latitude and longitude for a point inside a map can be roughly determined by the procedure described below. We will use as a worked example the location of Phantom Ranch on Topographic Map 1 (Bright Angel, Arizona).

For latitude, measure the distance of the point (Phantom Ranch) from the bottom edge of the map (9.1 cm). Next, measure the length of the map along a side from top to bottom (23.5 cm). Divide the first distance by the second (9.1/23.5 = 0.387). Determine the number of minutes of latitude spanned by the map (36°12' – 36°3' = 9') and then multiply these minutes by the earlier dividend (9.2 × 0.387 = 3.5'). Add the result to the latitude of the bottom edge of the map (36°3' + 3.5' = 36°6.5'). This sum is the latitude for the point of interest (36°6.5' for Phantom Ranch). Note that to simplify the calculations we are using fractional minutes instead of seconds (36°6.5' instead of 36°6'30", where 0.5 minutes equals 30 seconds; 0.5' × 60" = 30").

To find the longitude a similar calculation is done. Measure the distance of the point from the right edge of the map (8.0 cm) and then measure the width of the map along either the top or bottom (19.8 cm). Divide these two numbers (8.0 / 19.8 = 0.404) and multiply the result by the number of minutes of longitude spanning the map (0.404 × 9' = 3.6'; where 112°11' – 112°2' = 9'). Add this product to the longitude of the right/east edge of the map to obtain the longitude of the point of interest (112°2' + 3.6' = 112°5.6' or 112°5'36" W for Phantom Ranch).

Besides latitude and longitude, there is another type of coordinate system used on some USGS maps: **range** and **township**. These coordinates can be seen in portions of all maps in this manual (especially Map 3), and appear as black, numbered squares called **sections**. A section is one of the standard divisions of an area defined by a range and township, and normally measures 1 mile on a side.

Exercise 6B

All the questions that follow apply to Topographic Map 1 (Bright Angel, Arizona). Incidentally, the area shown can also be seen in the anaglyph on the classroom wall. You might first look at the anaglyph to get a "feeling" for the topography.

Write your answers on the sheet provided at the end of this chapter.

1. What is the maximum relief (to the nearest 100 feet) in the area south of the Colorado River?

    Maximum relief in feet: _____

2. What is the straight-line distance, in both miles and kilometers, between the summit (highest point) of the hill called Manu Temple (sector 4-A) and the site of Burro Spring (sector 3-E)?

    Distance in miles: _____

    Distance in kilometers: _____

3. (A) Which summit is higher: Bradley Point (sector 5-C/D) or Grandeur Point (sector 2-E)? (B) What is the elevation difference, in feet, between the two summits?

    (A) Highest summit: _____

    (B) Elevation difference in feet: _____

4. A popular hiking trail starts at Yaki Point on the South Rim of the Canyon (sector 4-E). This is known as the Kaibab Trail. It takes off, toward the north, following Cedar Ridge to O'Neill Butte, then continues along an unnamed ridge to Natural Arch, passes by The Tipoff, crosses the Colorado River at the Suspension Bridge, and then follows a creek to Phantom Ranch (sector 3-D) where it ends. (A) What is the distance, in miles, between Phantom Ranch and Yaki Point <u>along the trail</u>? (B) What is the <u>maximum</u> relief, in feet, along the trail? For Part A, you might find it easiest to mark off the distance for each trail segment consecutively along the edge of a sheet of paper, and then compare the total length of the trail against the graphic scale.

    (A) Distance in miles: _____

    (B) Relief in feet: _____

continued

5. What are the latitude and longitude of the summit of the butte called Isis Temple in sector 2-B (to the nearest tenth of minute)?

Latitude: _____

Longitude: _____

6. What is the gradient (in feet per mile) of Garden Creek (sectors 2E and 3D) between the head of its canyon by the road (i.e., just to the right of Section Number 23) and where it enters the Colorado River just below the 'resthouse'? For purposes of this calculation, use the straight-line distance between these two points. This is a reasonable approximation because the creek is relatively straight due to the fact that it follows a fault.

Gradient in feet per mile: _____

Hand in the filled-out answer sheet to your instructor. This will be graded and returned to you.

Exercise 6C

All the questions that follow apply to Topographic Map 2 (Maumee, Ohio).

Write your answers on the sheet provided at the end of this chapter.

1. Note the former river channel (meander scar) that passes to the north around the high ground starting near the word "Napoleon" (on the road) to the southwest in sector 1-D and returning to the modern channel just northeast of Turkeyfoot Rock in sector 3-B/C. How high, in feet, above the river level shown on the map must the Maumee rise during flood in order to flow through the meander scar, bearing in mind that the river flows toward the northeast?

    Flood height in feet: _____

2. The area shown on the map is the site of the famous "Fallen Timbers Battle." In 1794 the American army under General 'Mad' Anthony Wayne clashed with warriors of the Ottawa tribe under Chief Turkey Foot. The Indians lost and as a result Midwest was opened up for settlement by colonists from the east coast. The battle is thought to have started at the river near Turkeyfoot Rock and concluded on the higher ground just west of the cloverleaf highway interchange in sector 3/4-A. How much relief, in feet, is there between these two parts of the battlefield?

    Amount of relief in feet: _____

3. What are the latitude and longitude of the Fallen Timbers State Memorial in sector 3-B (the tiny circular drive just off highway US-24 on its south side)?

    Latitude: _____

    Longitude: _____

4. The proportional scale for the map is 1:24,000. (A) Convert this to a verbal scale: "one inch equals __?__ miles." (B) How many meters on the ground does one centimeter on this map represent?

    (A) "one inch equals _____ miles": _____

    (B) How many meters: _____

Hand in the filled-out answer sheet to your instructor. This will be graded and returned to you.

Chapter 7

Stream Hydrology

The Hydrologic Cycle

The hydrologic cycle describes the movement of water through all parts of the Earth's system. Water starts out as **precipitation**, as rainfall or snow. Precipitation is formed from the condensation of water vapor into ice crystals and water drops in the atmosphere. Some precipitation reaches the ground. Most precipitation falls directly on water bodies (i.e. the Oceans). Water that reaches the ground either **infiltrates** (soaks into the ground) or runs off. Water which infiltrates flows through the soil. It can be **evaporated** from the ground, returning to the atmosphere, or **transpired** through plants and is also returned to the atmosphere. What is not evaporated or transpired becomes **ground water**, which is the water present below the Earth's surface, filling the pore spaces and fractures in the subsurface. Water flows as ground water until it returns to the surface in stream, lakes or other bodies of water. This water joins with surface runoff, and flows towards a lake or ocean as **surface water flow**. Water is evaporated from these water bodies and returns to the atmosphere as water vapor.

Stream Properties

There are many ways to describe the size stream or river. One is by the length of a stream, from where it begins (the **headwaters**) to where it discharges into a lake or ocean (the **mouth**). Another way is to measure the **drainage basin**, the entire area which drains into a river system. Yet another is to measure the size of a channel at any given location along the length, the depth, the width.

A second way to measure the size of a river is to measure the amount of water which flows through the river. The **velocity** of a river is a measurement of the speed of flow through the river at a given location. **Discharge** is a measurement of the volume flowing in the river at a given point. The two are related by the area of the river cross section at that point (discharge = average velocity × cross section area). Discharge can increase by either increasing the velocity or by increasing the cross sectional area of a river. Velocity of a river is directly affected by the **slope** (also called **gradient**) of a stream. The slope is a measure of the vertical difference of the channel height divided by the channel length measured along the channel. A stream's gradient varies from steep to gentle along its course, being steepest near its head and gentlest toward its mouth.

A third way to discuss river size is to measure the amount of sediment that is carried in a river. River systems can carry material in three primary ways, **dissolved load**, **suspended load**, and **bed load**. Dissolved load is where material is dissolved into its constituent ions in the water and carried along in solution. Suspended load is the load that is carried along in the water as it moves. If the water is slowed or stopped, this material will settle out to the bottom of the water. Sediment which is moved along the bottom of the stream or river bed, being pushed or rolled along, but which is too heavy to be suspended in the water is called bed load. The **competence** of a stream describes how large the largest particles are that can be carried in suspended load. The more competent a river is, the larger the particle that is carried. Competence depends primarily on a river's velocity. The volume of sediment that a river can move is called the

FIGURE 7.1 The Hydrologic Cycle

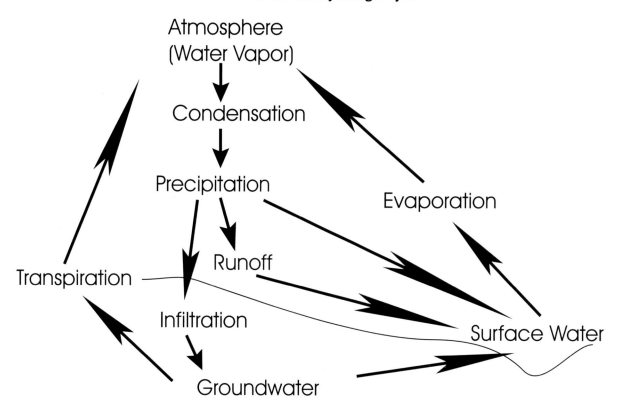

river's **capacity**. Capacity is a function of both velocity and discharge.

When a stream which is carrying sediment loses capacity by slowing down, it deposits sediment. Sediment deposited by streams is called **alluvium**. When a stream enters the ocean, the velocity slows down very quickly. This results in sediment being deposited where the channel meets the water. Such deposits are called **deltas**.

Channel Patterns

The velocity, sediment load and carrying capacity of a stream can cause a stream to develop a pattern different from the original channel. If you examine the Mississippi River, for example, you will see that the channel is not straight, but curves around inside of the valley. A sinuous channel which shifts across a valley is called a **meandering stream** (Figure 7.2). When sediment load is very

large, such as a pattern of a **braided stream** is formed, where the channel is split up into smaller channels, and the flow separates around bars in the stream.

In a meandering stream, flow velocity is not the same at all locations in the stream. This results in areas of the stream which will have higher capacities than others. As a result, some areas of the stream erode sediment, and others deposit sediment. A **cutbank** is a portion of the stream on the outside of a curve or **meander**. The velocity of the stream is highest at the cutbank, and as a result the stream can erode and carry more sediment. On the inside of a curve the velocity is slower, causing sediment deposition on a **point bar**. As a result, the channel migrates towards the outside of the curves in a stream, making the stream more sinuous as erosion progresses. If a meander curve is cut off by the main channel, it becomes an **oxbow lake**.

FIGURE 7.2 Channel Characteristics

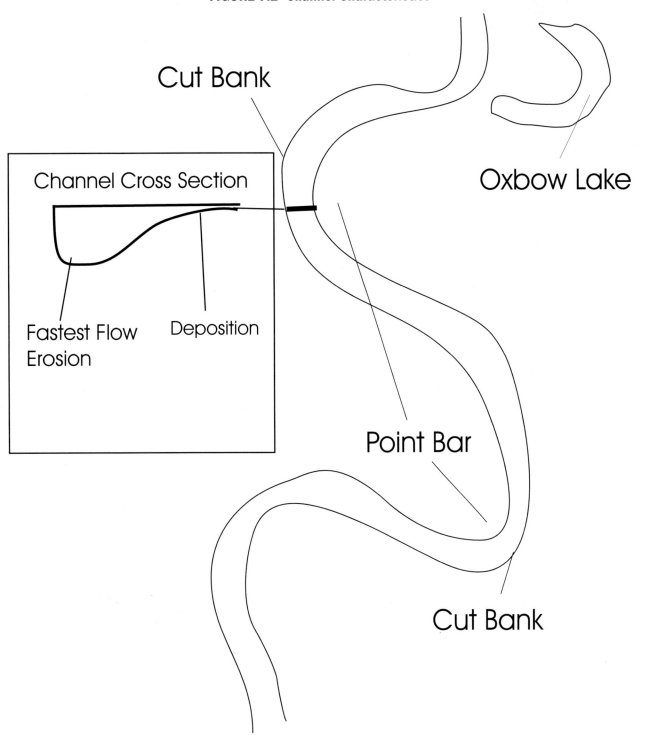

Cut Bank

Channel Cross Section

Fastest Flow
Erosion

Deposition

Oxbow Lake

Point Bar

Cut Bank

Exercise 7: Exploring the Changes in a Stream Channel

Introduction: This lab will attempt to simulate the conditions of a meandering stream. Since meandering streams can take thousands to millions of years to fully form, you will need to begin with some initial conditions to allow for our experiment to run to completion in a single lab session. It will be your responsibility to observe and record the important initial condition such that you can compare how they change through the course of lab and through different conditions of flow.

Scenario #1

1. Cut a straight channel in the sand parallel to the length of the stream table that will clearly channelize flow, but does not cut downward all the way through the sand until the lower end of the table is reached.

Question:

Predict what will happen in this scenario. Will there be erosion or deposition? Where? How do you think the stream bed will change?

Draw a picture of how you think the channel will look at the end of the experiment.

continued

2. Turn on the water pump and adjust clamp to give a smooth steady flow of water into the channel. Make note of how the sand moves, and sketch the stages as the stream develops.

3. Run the experiment about 25–30 minutes, or until a meander system is developed.

   Sketch the changes to the stream during the stages of development in scenario #1

Start

End

Questions:

How did the path of the stream change?

Why does this phenomenon occur?

What landform developed at the base where the stream entered the water? How did it evolve through time?

Were your predictions the same as or different than your observations? How would you modify your prediction of stream behavior based on your observations?

continued

Scenario #2

1. Cut a meandering channel with at least 2 full meanders from the source to the water.

2. Turn the water on to a slow, steady stream.

Question:

Predict what will happen in this scenario. Will there be erosion or deposition? Where? Will it be faster in some places than others? How do you think the stream bed will change?

3. Run the experiment for 25–30 minutes, or until the meanders have been significantly reshaped.

4. Using an eyedropper, streak some food dye across the stream near the head and watch it move downstream. Where is the water movement fastest? Slowest?

5. Observe the motion of individual sedimentary particles. Is the stream moving them by Dissolution, suspension or bedload?

6. Observe the changes in the meander loops. Where is erosion taking place?

7. Do you see places where the banks are slumping into the stream? Where is deposition taking place?

8. Do the places of erosion and deposition correlate with the places of fast and slow moving water?

Sketch the channel at the beginning and the end of your experiment. Note locations of fastest and slowest movement

Label your sketches where any of the following developed:

Cut Banks

Point Bars

Oxbow lakes

Meander scars

Along the length of the channel, where was the one place the greatest sediment was excavated. Why there?

How did sediment size affect transport?

Were your predictions the same as or different than your observations? How would you modify your prediction of stream behavior based on your observations?

continued

Scenario #3

1. Smooth out all of the sand until it forms a smooth surface sloping downhill

2. Turn on the water to a smooth gentle flow

Question:

Predict what will happen in this scenario. Will there be erosion or deposition? Where? How do you think the stream bed will change?

3. Continue to run the water for 25–30 minutes

What did the system look like while it evolved?

Sketch the channel shapes for 4 stages of the streams evolution

Were your predictions the same as or different than your observations? How would you modify your prediction of stream behavior based on your observations?

General Questions:

How does the stream table simulate the natural hydrologic cycle? What components of the natural system are missing?

Chapter 8

Geologic Structures and Geologic Maps

Note: *you will need a red pencil and a ruler for some of the exercises in this chapter.*

Introduction

Rocks in the Earth's crust are seldom left undisturbed after their initial formation. The same tectonic forces that are moving lithospheric plates around the globe are also applying **stress** to all parts of the crust. In places where the stresses are particularly great, such as in the mountain-building regions where lithospheric plates collide, the rocks are **deformed**. The different kinds of deformation are referred to as **geologic structures**.

Geologic maps show the geology underlying the Earth's surface. They do this by depicting two aspects of the rocks: the areal distribution of the different rock bodies (sedimentary layers, igneous intrusives, etc.) and the geologic structures (if any) imposed upon them. The maps only show what is exposed at the surface (that is, the **rock outcrops**). To illustrate the unseen geology below the surface, geologists draw **geologic cross-sections** which, like topographic profiles, are imaginary vertical cuts through the crust. The distribution and structure of rock bodies in the subsurface are deduced from the surface geology plus, where available, data from water and especially oil wells.

Tilted and Folded Rocks

Rocks that have been tilted or folded are said to be **plastically deformed** because, like plastic, they have been permanently warped into new shapes by the stresses applied to them. The deformation occurs, of course, in all rock types but is especially obvious in layered sedimentary rocks where the **attitude** (three-dimensional orientation) of the beds is evident.

The attitude of a given bed can be described in terms of its **direction of strike, direction of dip**, and **angle of dip** (Figure 8.1). The strike is an imaginary horizontal line drawn <u>on</u> the **bedding plane** (that is, the physical break between adjacent beds). It has a direction which can be measured with a compass, similar to what hikers use, and expressed as so many degrees and minutes east or west of geographic north. The dip is always perpendicular (at right angles) to the strike and corresponds to the compass direction in which the bed is inclined downward from the horizontal. The angle of dip is the angle between the horizontal and bedding planes.

On maps showing geologic structures, geologists indicate bedding attitudes with a **strike-and-dip symbol** (Figure 8.2). This "T" symbol is drawn on a map with the same geographic orientation as the sedimentary bed it represents. Thus, the top of the T is aligned with the strike direction and its stem is parallel with the dip direction.

Tilting, folding and other geologic structures are best illustrated using a **block diagram** like that in Figure 8.3. Such perspective diagrams show the three-dimensional aspects of a geologic structure by including a surface ("map") view and two vertical cross-sections.

Tilted rocks are inclined from their original (usually horizontal) position (like a tabletop that has been tipped) but are otherwise undeformed. Folded rocks can assume one of three basic forms: **monocline** (beds are bent in one direction

FIGURE 8.1 Strike and Dip of a Sedimentary Bed

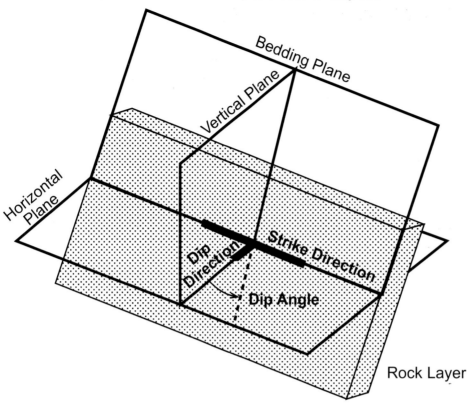

only), **anticline** (beds are bent convex upward), and **syncline** (beds are bent concave upward) (Figure 8.4). The latter two structures are also seen in Figure 8.3. Figure 8.5 shows the parts of a fold. The **axial plane** divides a fold into two **limbs** that dip in opposite directions. The imaginary line formed by the intersection of the axial plane and a bedding plane is called the fold's **axis** (like the crease in a folded sheet of paper). Fold axes are represented on geologic maps with the standard symbols shown in Figure 8.2.

Figure 8.6 shows a series of anticlines and synclines with axial planes (shown only for the anticlines) that vary from vertical (producing symmetrical folds; "A") and slightly inclined (producing asymmetrical folds; "B") to greatly inclined (producing an **overturned** fold; "C"). Note that the attitude of the beds on the right limb of the overturned anticline would be represented on a map with the special overturned strike-and-dip symbol in Figure 8.2.

Figures 8.3 to 8.6 show folds with horizontal axes. Fold axes, however, are often inclined (like

the crease in a sheet of paper that has been tilted from the horizontal). Such structures are referred to as **plunging folds** (Figure 8.7). In this figure, the stippled surface above the block diagram represents the form and position of a bedding plane prior to its removal by erosion.

Anticlines and synclines can plunge in opposite directions (see Figure 8.2 for symbols). Such **doubly plunging folds** have an elliptical outcrop pattern (Figure 8.8). As the outcrop pattern becomes more circular, such structures are referred to as **domes** and **basins** (Figure 8.9).

You will notice in Figures 8.8 and 8.9 that the attitude of the beds can be deduced on the geologic maps from the V-shaped bedding contacts where there are streams: a "V" points in the direction of bedding dip. If there is no V then the beds are vertical. In Figures 8.3 to 8.9 you will also notice that on the geologic maps the oldest beds are always exposed in the center of anticlines and domes, and the beds get progressively younger outwards. Just the opposite is true for synclines and basins.

Physical Geology Laboratory Exercises (EEES 1020)

FIGURE 8.2 Standard Symbols for Geologic Structures

STRIKE-AND-DIP SYMBOLS FOR BEDS

Bed is inclined X degrees from horizontal.

Bed is vertical.

Bed is overturned X degrees from horizontal.

Bed is horizontal.

FOLD AXES SYMBOLS

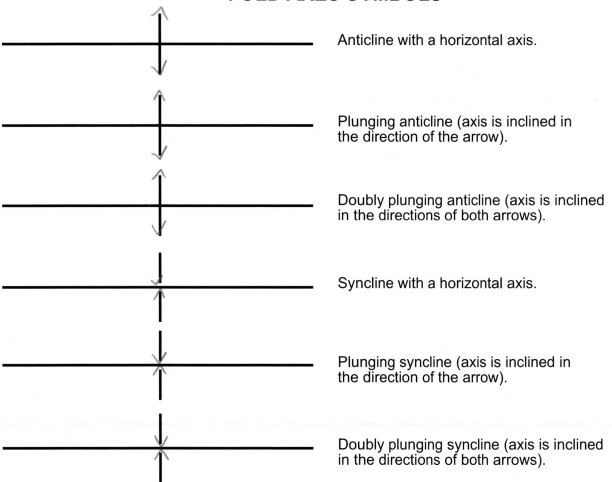

Anticline with a horizontal axis.

Plunging anticline (axis is inclined in the direction of the arrow).

Doubly plunging anticline (axis is inclined in the directions of both arrows).

Syncline with a horizontal axis.

Plunging syncline (axis is inclined in the direction of the arrow).

Doubly plunging syncline (axis is inclined in the directions of both arrows).

FIGURE 8.3 Example of a Block Diagram

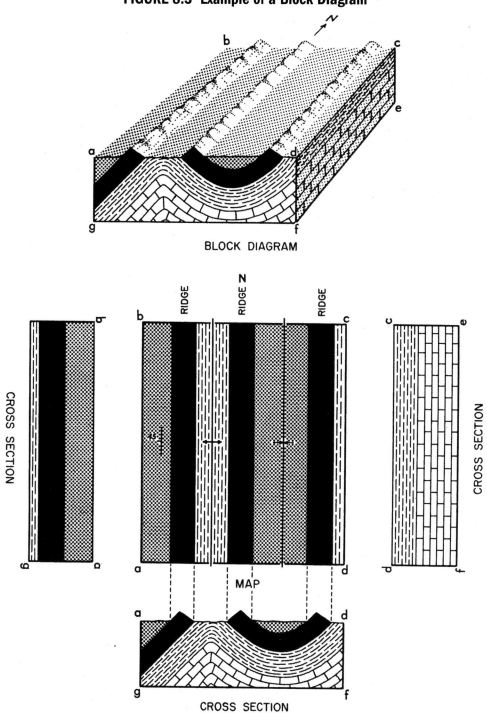

BLOCK DIAGRAM

MAP

CROSS SECTION

CROSS SECTION

CROSS SECTION

RIDGE RIDGE RIDGE

Figures from Laboratory Manual for Physical Geology by J. Zumberge, 1967. Reprinted by permission of The McGraw Hill Companies.

FIGURE 8.4 Basic Types of Folds

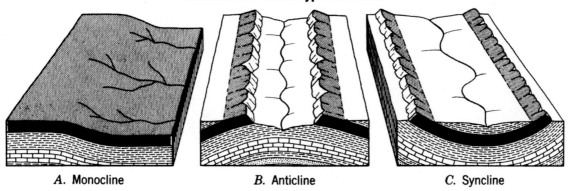

A. Monocline   B. Anticline   C. Syncline

Figures from Laboratory Manual for Physical Geology by J. Zumberge, 1967. Reprinted by permission of The McGraw Hill Companies.

FIGURE 8.5 Parts of a Fold

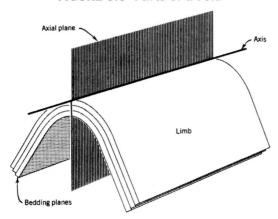

Figures from Laboratory Manual for Physical Geology by J. Zumberge, 1967. Reprinted by permission of The McGraw Hill Companies.

FIGURE 8.6 Variations in the Axial Planes of Folds

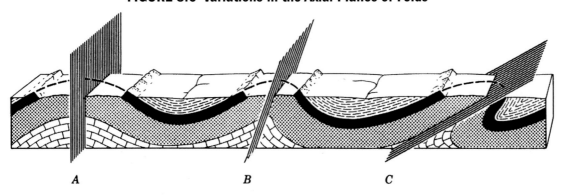

A   B   C

Figures from Laboratory Manual for Physical Geology by J. Zumberge, 1967. Reprinted by permission of The McGraw Hill Companies.

FIGURE 8.7 Plunging Folds

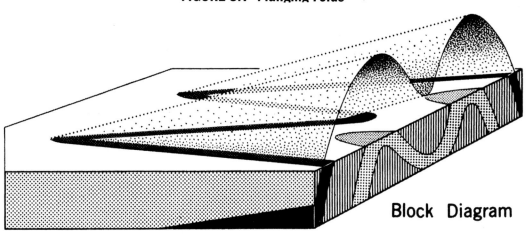

Block Diagram

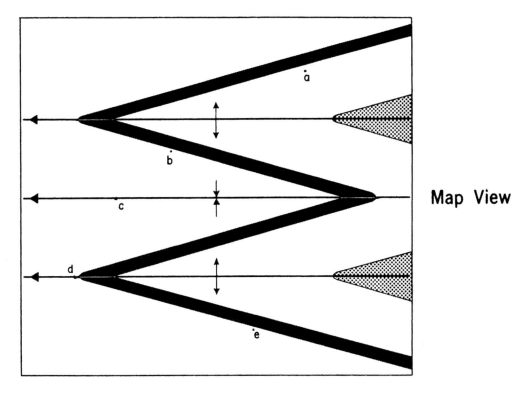

Map View

Figures from Laboratory Manual for Physical Geology by J. Zumberge, 1967. Reprinted by permission of The McGraw Hill Companies.

FIGURE 8.8 Doubly Plunging Folds

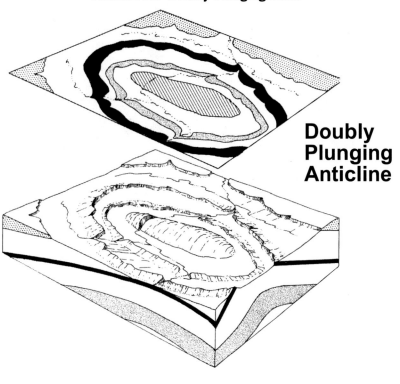

Doubly
Plunging
Anticline

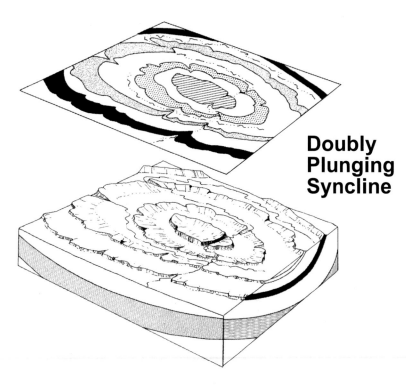

Doubly
Plunging
Syncline

Adapted from Physical Geology Lab Manual by Hamblin & Howard, 1967.

FIGURE 8.9 Domes and Basins

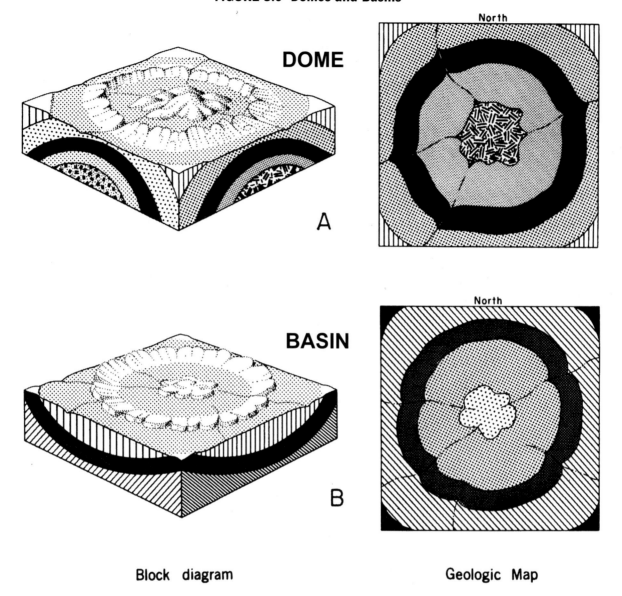

DOME

North

A

BASIN

North

B

Block diagram   Geologic Map

Figures from Laboratory Manual for Physical Geology by J. Zumberge, 1967. Reprinted by permission of The McGraw Hill Companies.

Fractures and Faults

Rocks sometimes break rather than fold, and in such cases they are said to have experienced **brittle deformation**. The simplest type of such deformation is the **fracture** (like a crack in a wall or pavement). It is common for a rock body to have numerous, roughly parallel fractures and these are referred to as **joints**. A fracture along which the rock on each side has moved in opposite directions is called a **fault**. Such movements result from the tectonic stresses applied to the rocks. The fault surface itself,

like a sedimentary bed, has an attitude that can be described by a strike and dip. There are, thus, **dip-slip faults** with movement parallel to the fault dip, and **strike-slip faults** with movement parallel to the strike (Figure 8.10). Dip-slip faults come in the **normal** and **reverse** varieties. In actuality, a given fault may experience more than one type of movement over its active life.

Every cubic mile of the Earth's crust is riddled with innumerable fractures and at least a few faults. This is true for both folded and tilted rocks as well as for non-deformed rocks. The vast

FIGURE 8.10 Basic Types of Faults

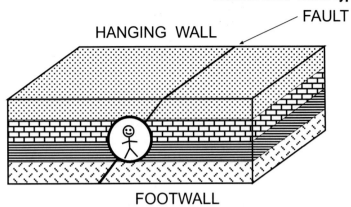

Definition of hanging wall and foot wall for a non-vertical fault: for a miner standing in a tunnel excavated along the fault, the hanging wall side would be "hanging" over his head and the footwall side would be under his "feet."

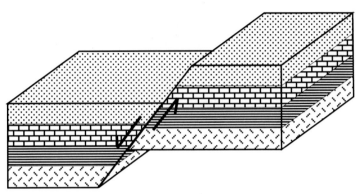

NORMAL DIP-SLIP FAULT: the footwall moves up relative to the hanging wall. All movement is vertical and parallel to the dip direction.

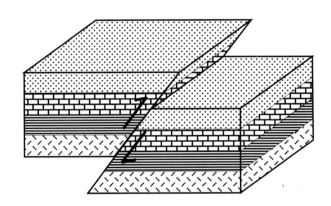

REVERSE DIP-SLIP FAULT: the hanging wall moves up relative to footwall. All movement is vertical and parallel to the dip direction.

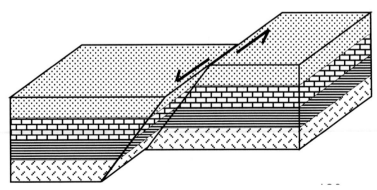

STRIKE-SLIP FAULT: all movement is horizontal (lateral) in the direction of the fault strike. The hanging wall and footwall do not move up or down relative to each other.

majority of these faults are "inactive" in that they have not experienced movement during the last 10,000 years (that is, during the Holocene epoch of the Quaternary period). It is the active faults that are responsible for most of the world's **earthquakes** (see Chapter 9).

Tectonic Stresses

The forces of deformation operating in the Earth's crust are of three types: **compression**, **extension** (also called "tension"), and **shear**. Compressional stress occurs where the crust is being "squeezed" such as in the regions near where lithospheric plates are colliding (that is, along "convergent plate boundaries"). Extensional stress occurs where the crust is being "stretched" such as in the regions where the lithospheric plates are being pulled apart (that is, in "rift zones" or along "divergent plate boundaries"). Shear stress occurs along the "transform-fault plate boundaries" where lithospheric plates are sliding past each other horizontally without either colliding or pulling apart.

The type of geologic structure formed in a rock is very much a function of the type of stress operating on it. The following relationships hold:

Type of Stress	Structure Formed
compressive	1. all anticlines and synclines
	2. all reverse faults
	3. some fractures
	4. some tilting, monoclines, domes and basins
extensional	1. all normal faults
	2. some fractures
	3. some tilting, monoclines and basins
shear	1. all strike-slip faults
	2. some fractures

Geologic Maps

A stated earlier, a geologic map shows the distribution of rock types and geologic structures in an area. In order to make them, geologists must spend a lot of time in the field describing the rocks, delineating their contacts and measuring their attitudes, and then plotting this information on topographic maps or aerial photographs. The maps and photographs are useful for more than just plotting field data. A careful examination of the topography seen in the maps and, especially, the photographs provides valuable information on rock bodies and geologic structures that might not be obvious to someone just walking around in the field. Many of the published geologic maps, especially those of the U. S. Geological Survey, also have topographic contours on them.

Exercise 8A

For the six block diagrams in Figure 8.11 draw correctly oriented dip-strike symbols beside the dark rock layer on top (the surface) of the block.

continued

FIGURE 8.11 Block Diagrams for Exercise 8A

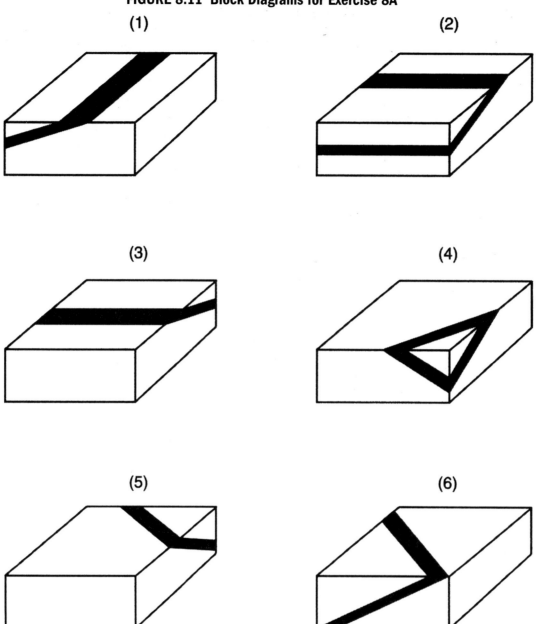

Figures from Physical Geology Exercises, 2nd edition by Jennifer Hinds. Copyright © 2004 by Kendall Hunt Publishing Company. Reprinted by permission.

Exercise 8B

Draw in the bedding contacts in the cross-section panel for the geologic map in Figure 8.12a. Also in this panel, number the sedimentary beds consecutively from 1 to 4, with 1 being the oldest and 4 the youngest.

FIGURE 8.12a Geologic Map for Exercise 8B

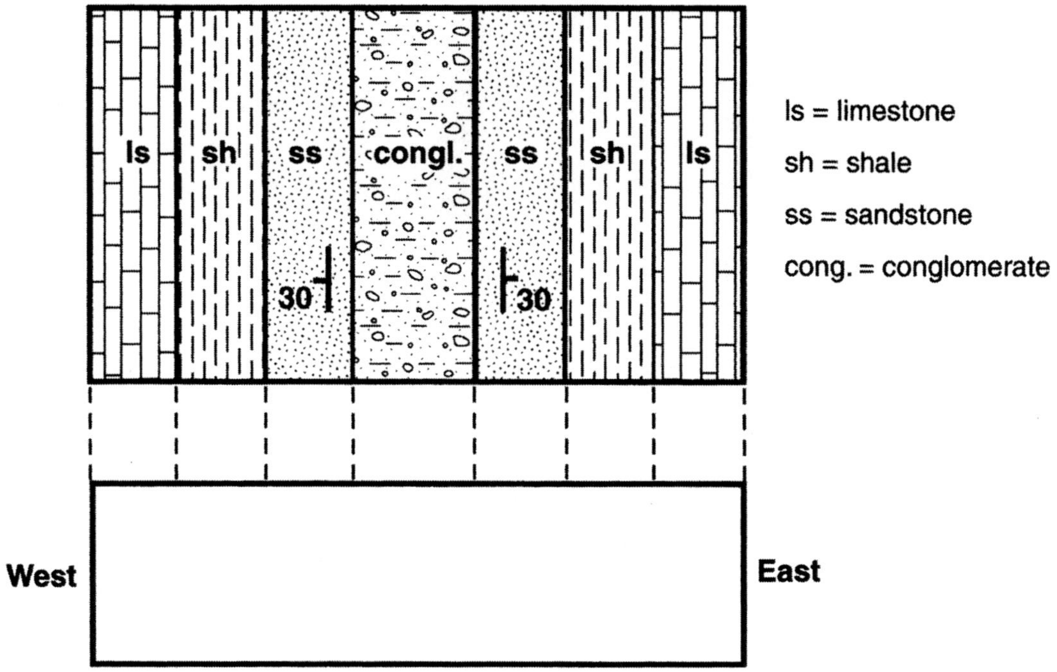

ls
sh
ss
congl.
ss
sh
ls
30
30
ls = limestone
sh = shale
ss = sandstone
cong. = conglomerate
West
East

Exercise 8C

Draw the bedding contacts on the top (surface) of the block diagram in Figure 8.12b.
Also draw in the correct fold symbol from Figure 8.2.

FIGURE 8.12b Block Diagram for Exercise 8C

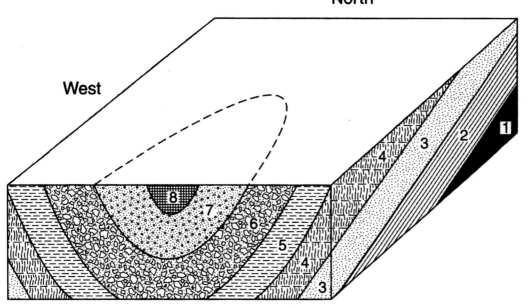

Exercise 8D

Complete the two cross-sectional views on each of the four block diagrams in Figure 8.13a. The relative ages of the sedimentary beds are indicated by the consecutive numbers with 1 being the oldest and the highest number being the youngest. Four blocks 3 and 4, you will need to apply the "rule of V's" (see contouring rule 8 in Chapter 6).

FIGURE 8.13a Block Diagrams for Exercise 8D

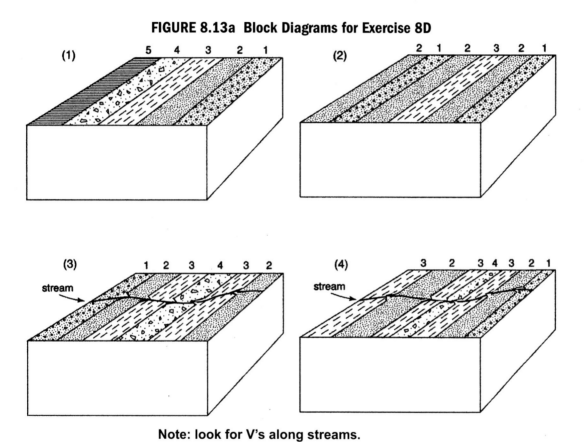

Note: look for V's along streams.

Exercise 8E

Draw in the bedding contacts on the blank sides and tops of the two block diagrams in Figure 8.13b. The blank portion of each block represents the up-thrown side of a fault. Also add the correct fold symbol from Figure 8.2.

FIGURE 8.13b Block Diagrams for Exercise 8E

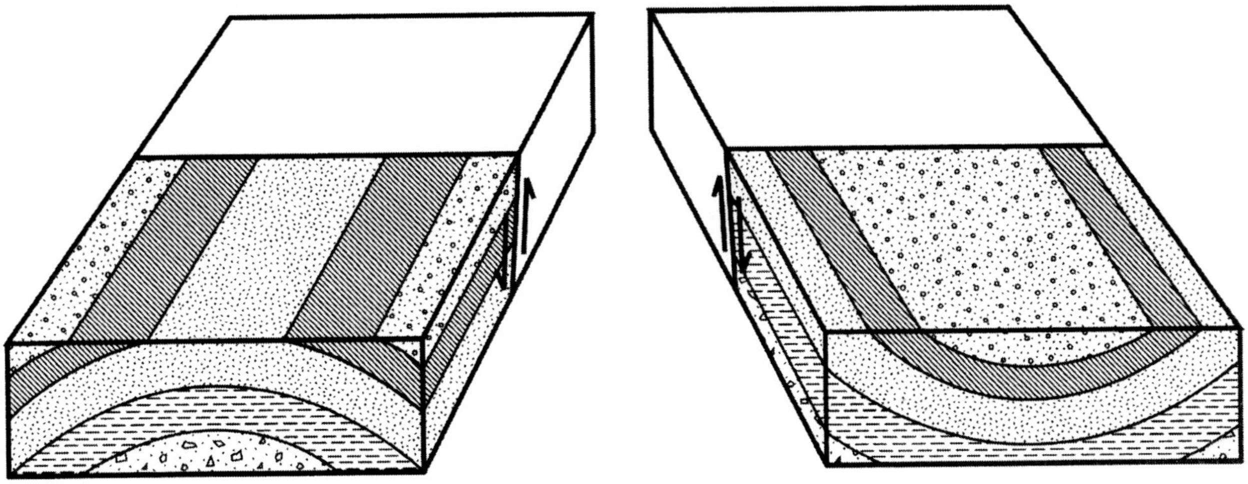

Exercise 8F

Figure 8.14a is the geologic map of the Black Hills (South Dakota). In Figure 8.14b draw the two geologic cross sections corresponding to the surface geology. The cross sections are for the lines represented by the left edge of the upper panel and the top edge of the lower panel. Also on Figure 8.14b, draw the correct fold symbol from Figure 8.2. Note that the geologic age is given for each rock formation and, as seen by comparison with Table 5.1, these are arranged from youngest at the top to oldest at the bottom.

continued

FIGURE 8.14a Geologic Map of the Black Hills, South Dakota (for Exercise 8F)

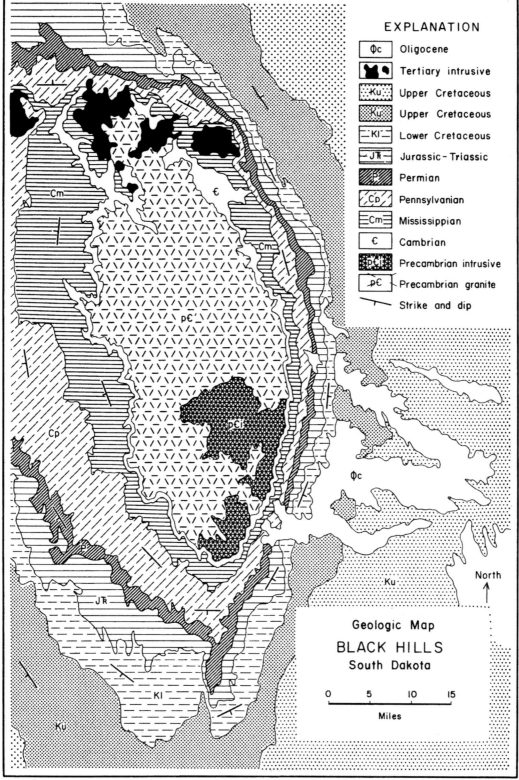

EXPLANATION

Фc	Oligocene
	Tertiary intrusive
Ku	Upper Cretaceous
Ku	Upper Cretaceous
Kl	Lower Cretaceous
JŦ	Jurassic–Triassic
P	Permian
Cp	Pennsylvanian
Cm	Mississippian
€	Cambrian
pЄi	Precambrian intrusive
pЄ	Precambrian granite
	Strike and dip

Geologic Map
BLACK HILLS
South Dakota

0   5   10   15
Miles

North ↑

Figures from Laboratory Manual for Physical Geology by J. Zumberge, 1967. Reprinted by permission of The McGraw Hill Companies.

116 Physical Geology Laboratory Exercises (EEES 1020)

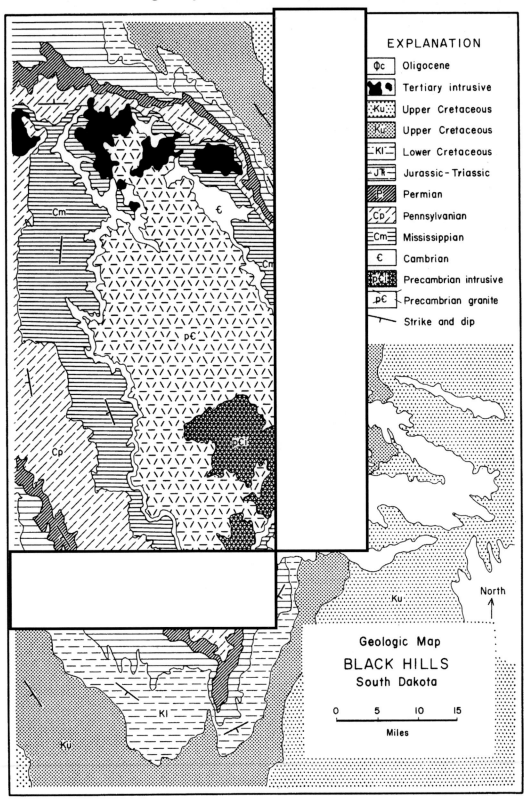

EXPLANATION

- Oc — Oligocene
- Tertiary intrusive
- Ku — Upper Cretaceous
- Ku — Upper Cretaceous
- Kl — Lower Cretaceous
- JR — Jurassic-Triassic
- P — Permian
- Cp — Pennsylvanian
- Cm — Mississippian
- € — Cambrian
- p€i — Precambrian intrusive
- p€ — Precambrian granite
- Strike and dip

Geologic Map
BLACK HILLS
South Dakota

0   5   10   15
Miles

North

Figures from Laboratory Manual for Physical Geology by J. Zumberge, 1967. Reprinted by permission of The McGraw Hill Companies.

Exercise 8G

Using the geologic map of Ohio in Figure. 8.15, deduce the nature of the geologic structure from the outcrop pattern of sedimentary rocks. Draw correctly oriented strike-and-dip symbols at the six solid dots scattered around the map, and also draw in any other appropriate structural symbols from Figure 8.2. Note that the patterns indicate the geologic ages of the sedimentary rock units.

continued

FIGURE 8.15 Geologic Map of Ohio (for Exercise 8G)

FIGURE 8.15 Geologic Map of Ohio (for Exercise 8G)

PERMIAN

PENNSYLVANIAN

MISSISSIPPIAN

DEVONIAN

SILURIAN

ORDOVICIAN

N

40 MILES

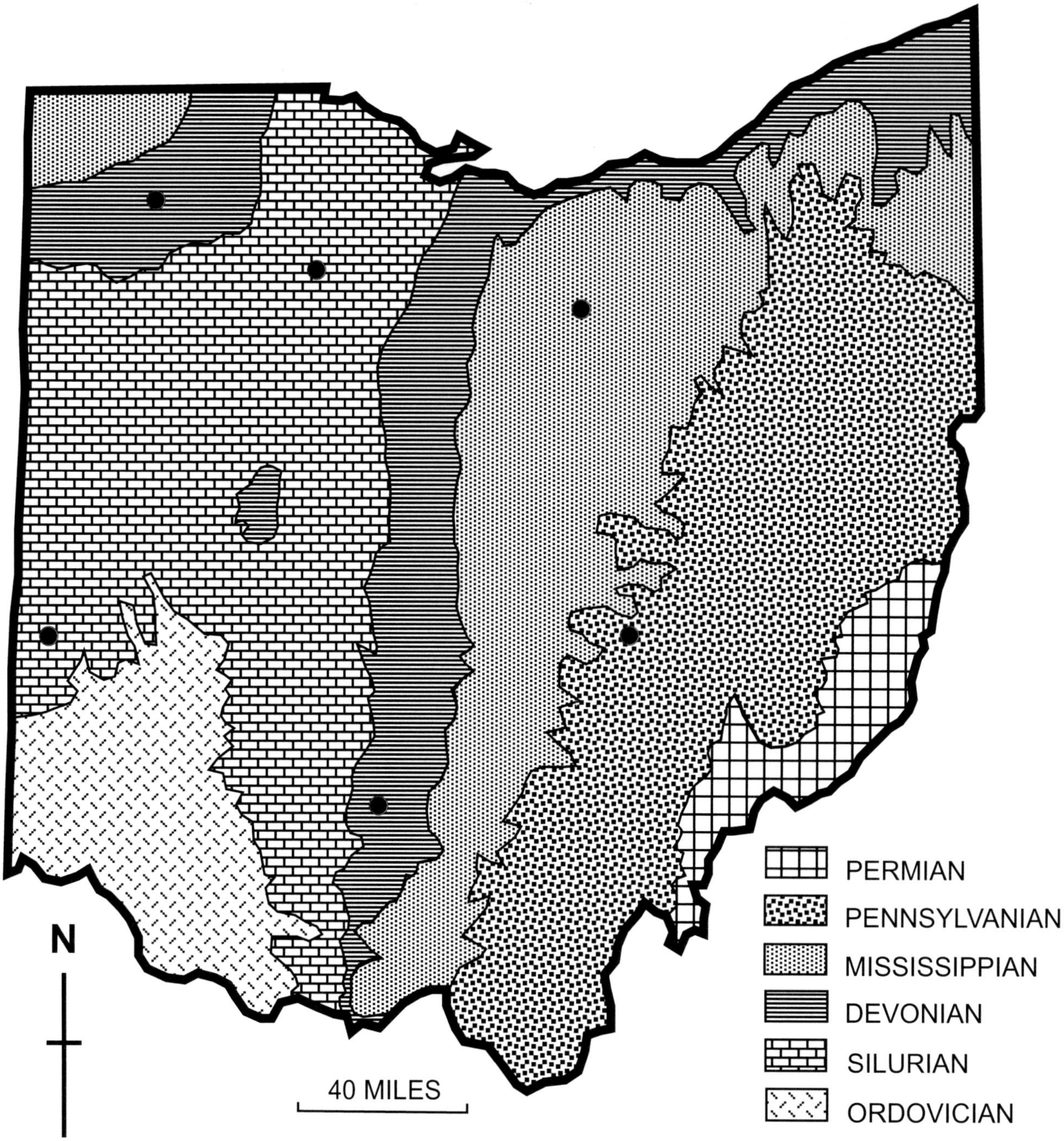

Exercise 8H

Examine the outcrop pattern of sedimentary rocks in the geologic map of Michigan (Figure 8.16a). What geologic structure is evident from this pattern? Write your response in the space provided on the answer sheet in Figure 8.16b. Also in Figure 8.16b, sketch a generalized geologic cross section along line joining Toledo, Ohio and Marquette, Michigan.

Hand in the block diagrams, maps, cross sections, and answer sheet for the above exercises. These will be graded and returned to you.

continued

FIGURE 8.16a Geologic Map of Michigan (for Exercise 8H)

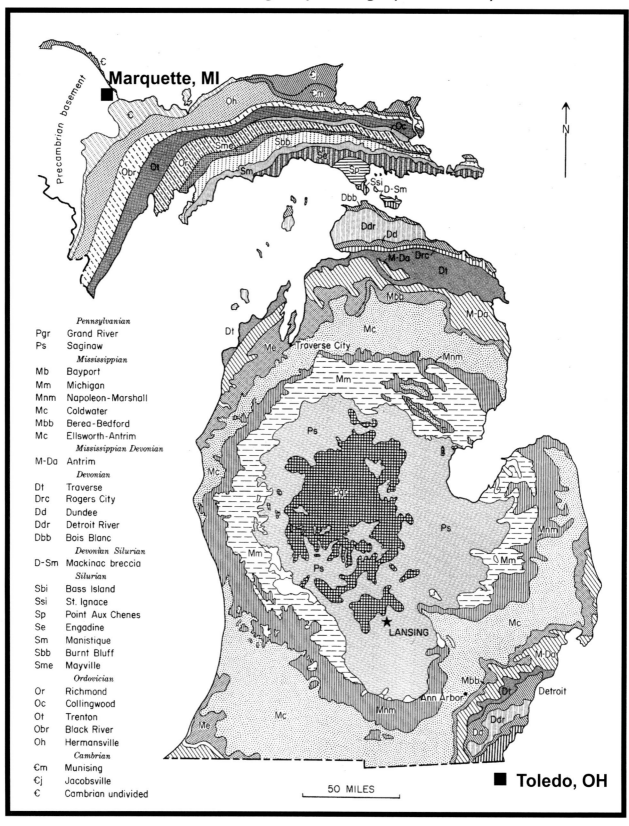

Pennsylvanian
Pgr Grand River
Ps   Saginaw

Mississippian
Mb   Bayport
Mm   Michigan
Mnm Napoleon-Marshall
Mc   Coldwater
Mbb Berea-Bedford
Mc   Ellsworth-Antrim

Mississippian Devonian
M-Da Antrim

Devonian
Dt   Traverse
Drc Rogers City
Dd   Dundee
Ddr Detroit River
Dbb Bois Blanc

Devonian Silurian
D-Sm Mackinac breccia

Silurian
Sbi Bass Island
Ssi St. Ignace
Sp   Point Aux Chenes
Se   Engadine
Sm   Manistique
Sbb Burnt Bluff
Sme Mayville

Ordovician
Or   Richmond
Oc   Collingwood
Ot   Trenton
Obr Black River
Oh   Hermansville

Cambrian
Ꞓm   Munising
Ꞓj   Jacobsville
Ꞓ   Cambrian undivided

Marquette, MI

Toledo, OH

50 MILES

FIGURE 8.16b Answer Sheet for Exercise 8H

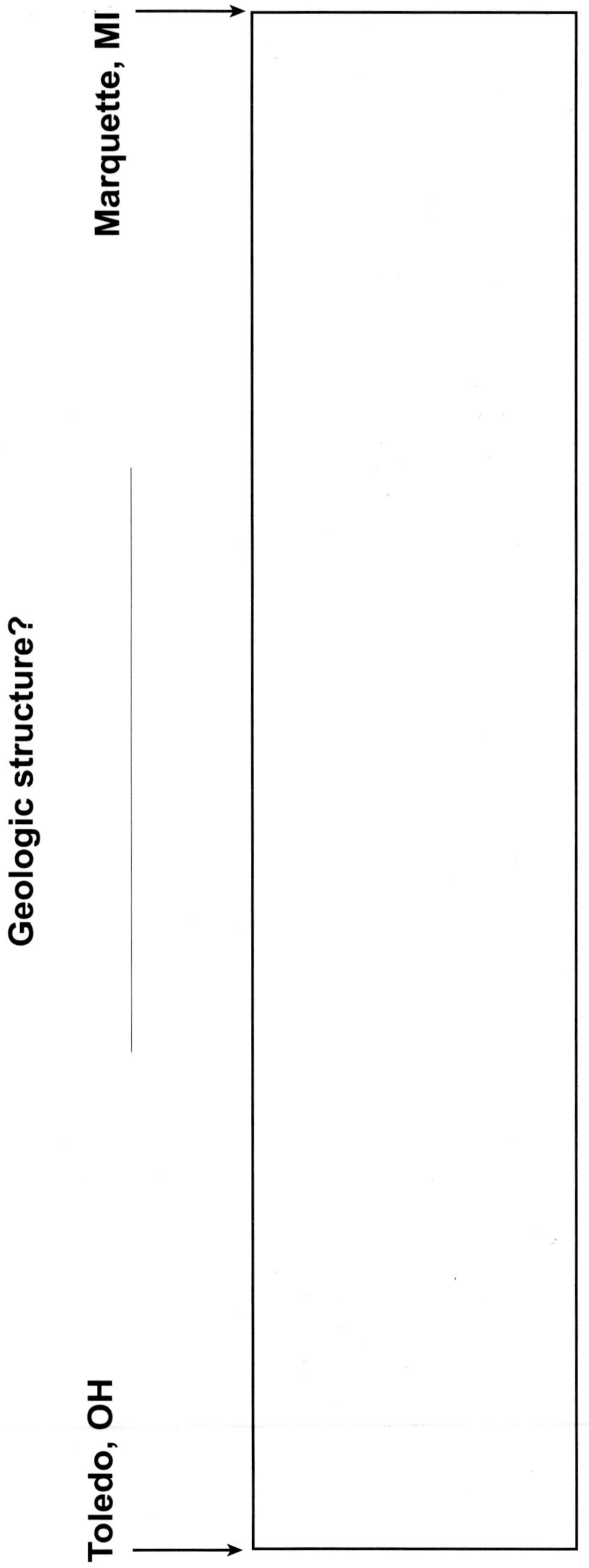

Geologic structure?

Marquette, MI

Toledo, OH

Chapter 9

Earthquake Analysis

Note: *a metric ruler and a drafting compass would be useful for the exercises in this chapter, but are not essential.*

Introduction

An earthquake results when the two sides of a fault try to move past each other but cannot do so because of **frictional resistance** along the fault plane. Such "locked" faults experience **elastic deformation** of the rocks on each side. Unlike permanent plastic and brittle deformations (as in folds and joints, respectively), elastic deformation is only temporary. When the deforming stress is removed, the elastic deformation disappears (like stretching [elastically deforming] a rubber band that resumes its original shape when it is released).

For active faults, the elastic deformation increases through time if the tectonic stress is continuously applied (like progressively stretching a rubber band). At some point the accumulated elastic energy along the fault will overcome the frictional resistance to movement, and the two sides of the fault will suddenly snap past each other and, in the process, remove the elastic deformation (like quickly releasing a stretched rubber band). The sudden movement sends out vibrations (**seismic waves**) through the surrounding crust and these are experienced as **earthquakes**. The foregoing explanation is known as the **Elastic Rebound Theory of Earthquakes**.

Earthquake Swarms

Slippage along faults almost never produces a single earthquake but rather a swarm of them (sometimes numbering in the thousands). In general, the more elastic energy released, the more earthquakes there are in the swarm. The **principal shock** is the most powerful earthquake in the swarm and is usually the first one felt. **Foreshocks** are earthquakes preceding the principal shock. These are usually very weak and so generally unfelt, and do not always occur. They are caused by incipient slippage along the fault and can happen hours or days before the principal shock. **Aftershocks** are earthquakes following after the principal shock and always occur! They are very common and can be anywhere from very weak to nearly as powerful as the principal shock. Aftershocks are caused by additional slippage along the fault as well as by slippage along nearby reactivated faults that were destabilized by the principal shock. They can occur for days, weeks or even months after the principal shock.

Focus and Epicenter

The **focus** of an earthquake is the place beneath the earth's surface where the earthquake starts (that is, the point where the initial fault rupture occurs). The depth of the focus is often reported in news accounts of earthquakes. For earthquakes of equal strength, the shallower the focus, the more damage is done at the surface by the ground shaking. This is true because an earthquake's energy is attenuated with increasing distance from the focus.

The **epicenter** is the point on the earth's surface directly above the focus, and is commonly thought of as "ground zero" for an earthquake because most of the damage will occur in areas at or close to the epicenter.

Seismic Waves and Seismo-Graphs

Whenever an earthquake occurs, the seismic waves propagate outwards from the focus in all directions. Some travel only through the Earth's interior and are called **body waves**, while **surface waves** can only travel across the Earth's exterior.

The body waves are of two types. The **primary** (**P** or compressional) waves are equivalent to sound waves, can travel through any medium (solid, liquid or gas), and propagate by alternately compressing and extending the medium through which they move in the direction of advance. In contrast, the **secondary** (**S** or shear) waves can travel only through solids, and propagate by causing the medium through which they move to vibrate perpendicular to the direction of advance (like whipping an electric cord or garden hose from side to side).

The two principal types of surface waves are the **Love**, which produces horizontal ground displacements with a motion like a sidewinder snake [note: this is not the same thing as the 'love waves' produced by adoring fans are sporting events!], and the **Rayleigh**, which produces rolling ground displacements causing the surface to visibly ripple rather like backward-rotating ocean waves.

Secondary waves cause most of the damage from ground shaking, both near the epicenter and beyond it, with a small portion of the damage near the epicenter also produced by the surface waves. The body and surface waves vary greatly in their velocities. The primary or P waves are the fastest with an average velocity of 16,000 mph whereas the secondary or S waves lag well behind at about 7,500 mph. In general, the P waves travel 1.8 times faster than the S waves. The surface waves are the slowest of all, but still manage a respectable 3,250 mph.

An instrument known as a **seismograph** detects the passage of all the seismic waves, and these are recorded by a pen dragging along a strip of paper on a rotating drum. This recording is known as a **seismogram** (see Figure 9.1). The arrival of each type of wave causes the pen to swing back and forth across the straight center line of the seismogram by a distance that is proportional to the amount of ground motion caused by the wave's passage (that is, its **amplitude**). The different arrival times for the various waves (P first, S second, and surface last) are also recorded on the seismogram.

Earthquake Strength

There are a variety of measures used to characterize the strength of an earthquake. We will consider here only the simplest of these, the **Richter magnitude**. This is a measure of both the

FIGURE 9.1 Example of a Seismogram

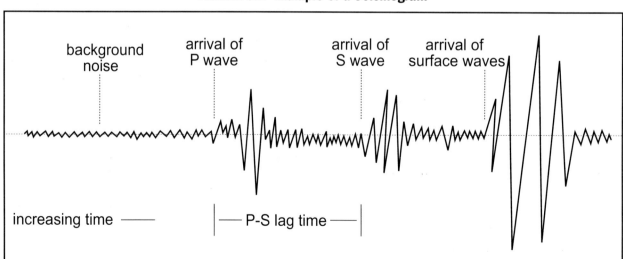

FIGURE 9.2 Nomogram for Determinig Richter Magnitude

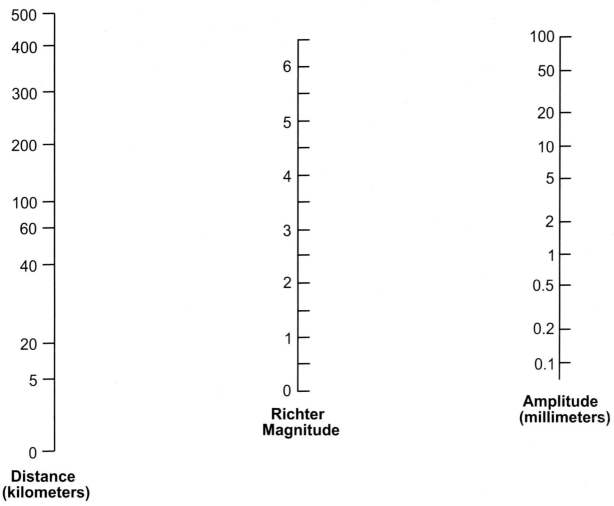

amount of ground motion and the elastic energy released, and these are deduced from the amplitude of the seismic waves as recorded on a seismogram. The Richter magnitude is proportional to the common (base 10) logarithm of the maximum width of the seismogram tracing for the S waves. There is much more to its calculation but we need not concern ourselves with it here. Happily, the calculations can be largely avoided altogether if one uses a nomogram like that in Figure 9.2, which is based on seismic wave amplitude and epicenter distance. The important point is that, being measured on a logarithmic scale, each 10-fold increase in seismogram tracing width (and, hence, ground motion) corresponds to an increase of 1 in Richter magnitude. The magnitude scale ranges from 0 upwards. Although it is open-ended, 10 is about the practical upper limit for the strength

of an earthquake. Rocks are not strong enough to accumulate the amount of elastic deformation needed to make an earthquake over 10. Each increase in magnitude by 1, corresponds to about a 30-fold increase in the amount of elastic energy released by the earthquake. Thus a magnitude 6 earthquake releases 30 times more energy than a magnitude 5,900 (30 x 30) times more than a magnitude 4, and 27,000 (30 x 30 x 30) times more than a magnitude 3.

Locating Earthquakes

Whenever a strong earthquake occurs in the world, people want to know three things about it: (1) the magnitude, (2) the distance to the epicenter, and (3) the precise geographic location of the epicenter. Answers to all of these questions are provided by the seismograms from one or more widely separated seismograph stations.

To find the distance to an epicenter, only one seismograph station is needed. One simply uses the lag in the arrival times for the P and S waves. The longer the lag, the greater the distance. This relationship is graphically depicted by a **travel-time plot** (see Figure 9.3). To find the latitude and longitude of the epicenter, at least three seismograph stations are needed, preferably widely separated and on different sides of the epicenter. In practice one would take a map and first plot the locations of the three seismograph stations. Next, using a drafting compass, one would then draw circles around each station with radii corresponding to the epicenter distances (based on P-S travel-time lag) from the stations. The epicenter is located where the three circles intersect.

FIGURE 9.3 Travel-Time Plot for P and S Seismic Waves

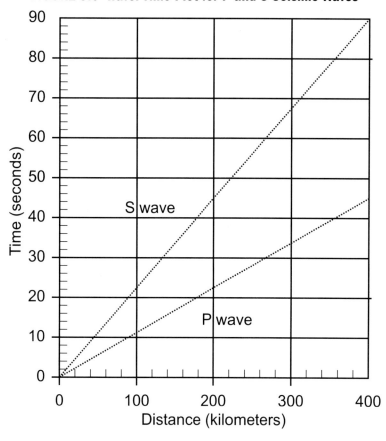

Physical Geology Laboratory Exercises (EEES 1020)

Exercise 9

PART 1: Look at the three seismograms in Figure 9.4. These come from seismograph stations in the three cities indicated and all record the same earthquake, which occurred in the Pacific Northwest region. On the answer sheet provided at the end of the chapter, fill in for each station the P-wave arrival time, S-wave arrival time, lag time between the P and S waves, and the maximum amplitude of the S waves.

The arrival times can be read directly from the seismograms. The timing marks along the bottom of each of these are subdivided into 5-second intervals (each 5 mm wide) with the clock times indicated every 30 seconds. For example, 4:33:30 means 4 hours, 33 minutes, 30 seconds. Estimate the arrival times (the first swing of the pen on the left side of each tracing) to the nearest second (or 1 mm). To find the maximum amplitude of the S waves, draw a straight, horizontal line from the highest part of a S wave tracing to the left side of the seismogram, which is subdivided into 10 mm increments. If you do not have a metric ruler for these measurements, you can use the one printed in Table 6.1.

PART 1

	P wave arrival time	S wave arrival time	Lag time between P & S waves	Maximum S wave amplitude
Ellensburg	_____	_____	_____	_____
Portland	_____	_____	_____	_____
Bellingham	_____	_____	_____	_____

continued

FIGURE 9.4 Seismograms for Earthquake in the Pacific Northwest

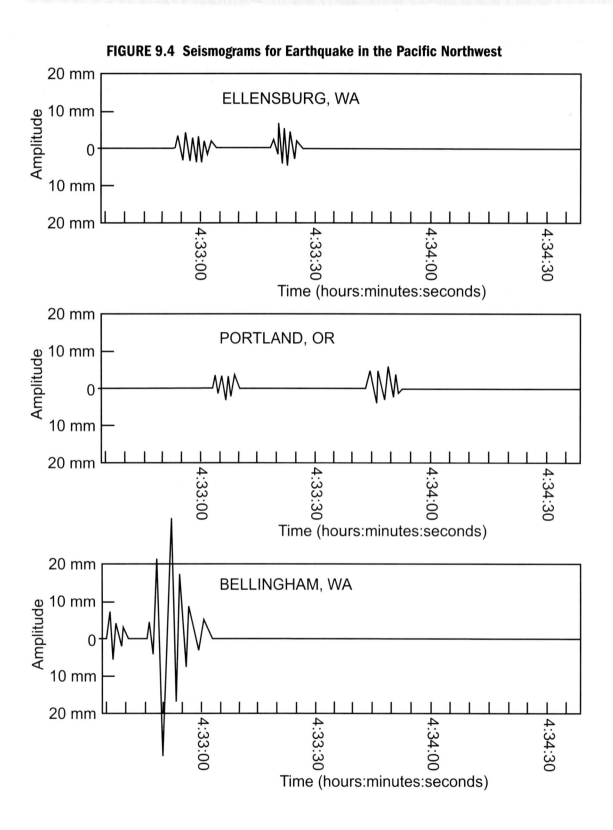

PART 2: Using the travel-time plot in Figure 9.3, determine the distance from each seismograph station to the epicenter and record it on the answer sheet. To do this for one of the seismograms, place the edge of a sheet of paper along the left margin of Figure 9.3 and mark off the distance corresponding to the lag time between the P and S waves. Next, again using Figure 9.3, orient the edge of the marked sheet vertically and slide it along the P-S travel time curves until the spacing of two marks exactly matches the distance between the curves. From where this occurs, read straight down to find the distance to the epicenter. For example, if the time lag on a seismogram is 32 seconds, place marks on the sheet edge at 0 and 32 seconds using the left side of Figure 9.3. This will correspond to an epicenter distance of 280 km.

PART 2

Distance from epicenter

Ellensburg   _____

Portland   _____

Bellingham   _____

continued

Exercise 9, continued

PART 3: Using the nomogram in Figure 9.2, find the Richter magnitude for the earthquake and record it on the answer sheet. Use the epicenter distance and maximum S-wave amplitude for the <u>closest</u> seismograph station. Mark the distance and amplitude on the left and right sides of the nomogram, respectively, and then draw a straight line connecting the two points. The point where the line crosses the middle scale corresponds to the Richter magnitude of the earthquake. Note that the left and right scales are logarithmic and so care should be taken in interpolating between the numbered tick marks.

PART 3

Richter magnitude = _____

PART 4: Plot the location of the epicenter on the map in Figure 9.5. To do this you need to draw circles around each of the seismograph stations with radii equal to the distances to the epicenter as determined in PART 2. The easiest way to do this is to use a drafting compass. If you do not have one, then use a sheet of paper with two <u>tiny</u> (pin) holes separated by the required distance. Using a pen or pencil, anchor one of the holes on a seismograph station and then placing another pen or pencil in the other hole, rotate the latter around the former. Where the three circles intersect is the location of the epicenter.

Hand in the map and answer sheet for the above exercise. These will be graded and returned to you.

FIGURE 9.5 Map of the Pacific Nothwest Showing the Locations of the Three Seismograph Stations

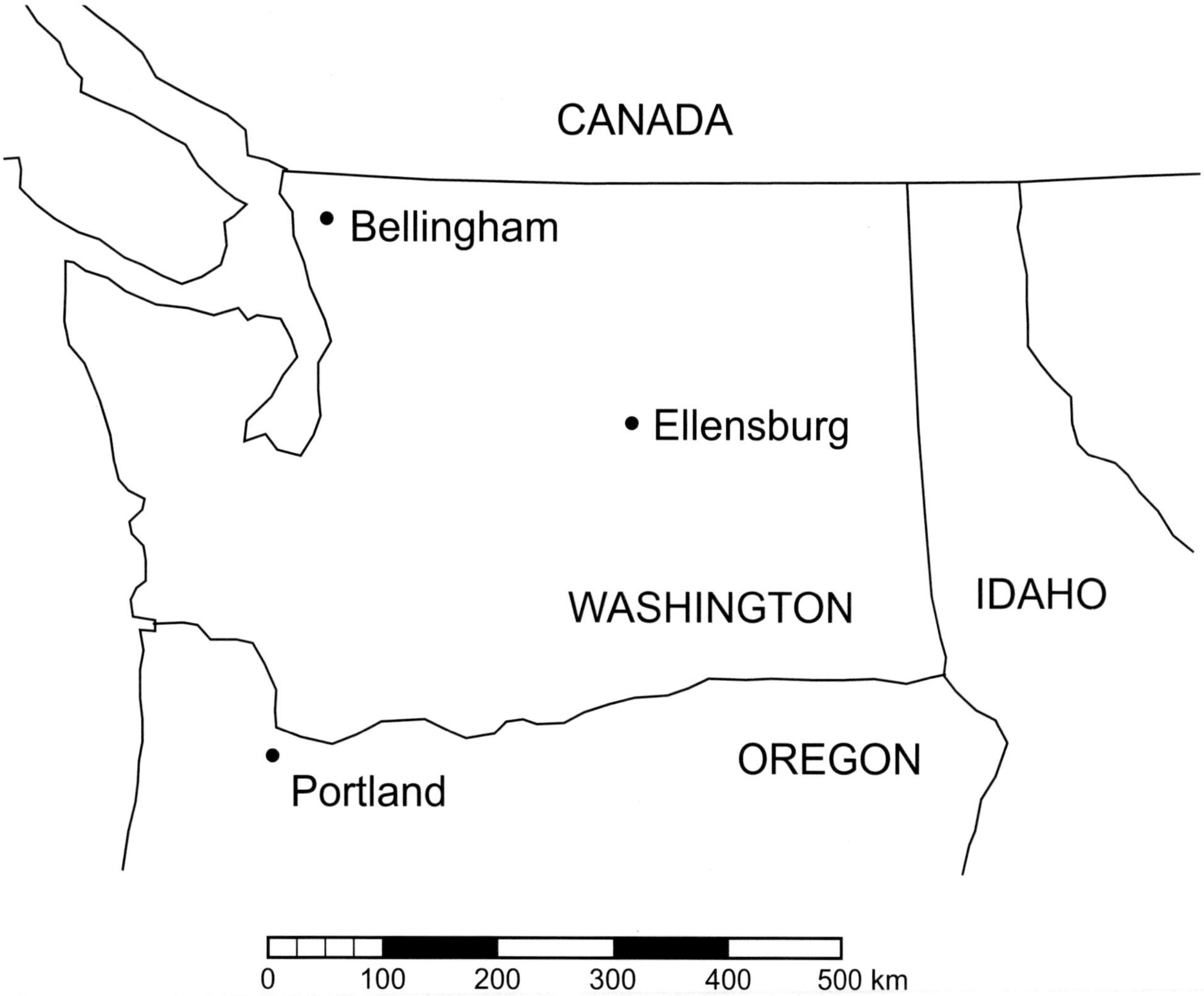

Exercise 9B

In this exercise you will use an experimental set up to investigate the nature of forces and motion in earthquakes.

For this experiment you will set up a board with a winch at one end. Connected to the winch by a cable and surgical tubing will be a cinder block. The cinder block rests on sandpaper on the board.

Make a prediction of what will happen when the crank is turned? Describe in detail what will happen. Will the block stay in place or move? Will it move smoothly or in short motions with pauses between? Is energy stored or released anywhere in the system?

Slowly turn the crank. Write down what you observe. Record the length of any movements that you observe.

How did your observations differ from your prediction? How would you modify your hypothesis to reflect the results of your experiment?

FIGURE 9.6 Equipment Diagram

continued

Reset the experiment. Place the second cinder block on top of the first.

Predict what you expect to happen. Will this be the same as the first experiment, or different? What differences do you expect?

Why do you expect these differences?

Slowly turn the crank. Write down what you observe. Record the length of any movements that you observe.

Were there any differences between the first and second experiment? If so, what were they?

What caused these differences? (Think about what forces have changed and if the amount of energy stored is different)

How does this model represent what may occur in the Earth's subsurface?

What does adding a second block in the experiment represent in the Earth's subsurface?

Chapter 10

Comprehensive Exercise

The purpose of this chapter is to give you the opportunity to pull together everything you learned in the previous chapters and apply it in a single comprehensive exercise. For the same geographic area, you will be identifying the rock types present, contouring the topography, making a geologic map, producing a combination topographic profile and geologic cross section, and interpreting the landscape development and geologic history.

FIGURE 10.1 Common Rock Patterns used in Geologic Maps and Cross Sections

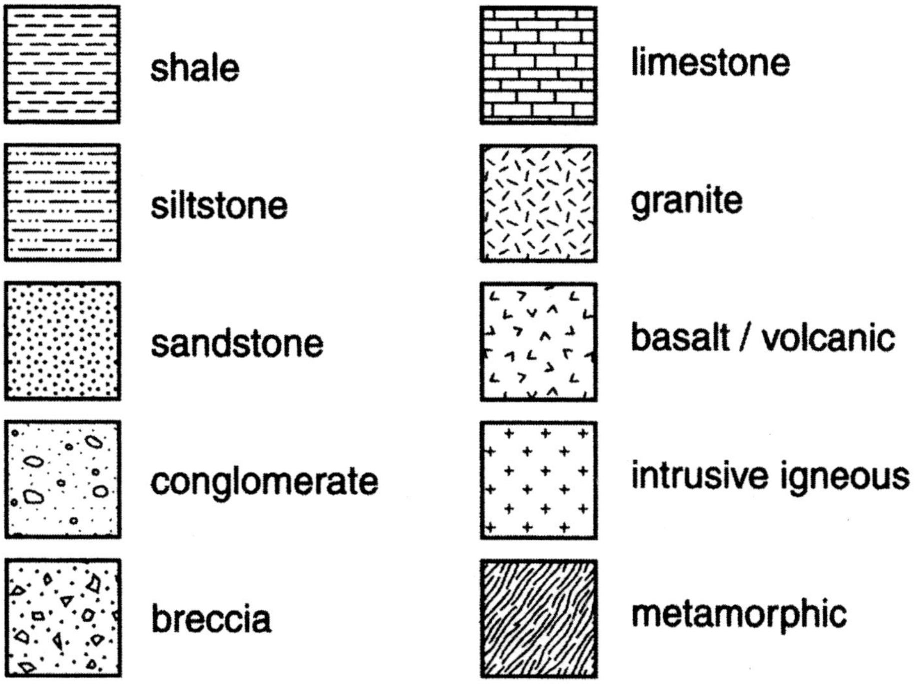

- shale
- siltstone
- sandstone
- conglomerate
- breccia
- limestone
- granite
- basalt / volcanic
- intrusive igneous
- metamorphic

Figures from Physical Geology Exercises, 2nd edition by Jennifer Hinds. Copyright © 2004 by Kendall Hunt Publishing Company. Reprinted by permission.

PART 1: Using a contour interval of 40 feet, complete the topographic contours in Figure 10.2 based on the spot elevations given. To get you started, the 760 foot contour is provided. Be sure to number all the contours you draw.

PART 2: Using the locations of rock units A, B, C, D and E, along with their corresponding strike and dip measurements, draw the approximate geologic contacts and indicate the geologic structure using the appropriate symbol from Figure 8.2. A fault crosses the northwest corner of the map with the up-thrown and down-thrown sides indicated by U and D, respectively. Unit C is an intrusive igneous dike.

FIGURE 10.2 Base Map for Topography

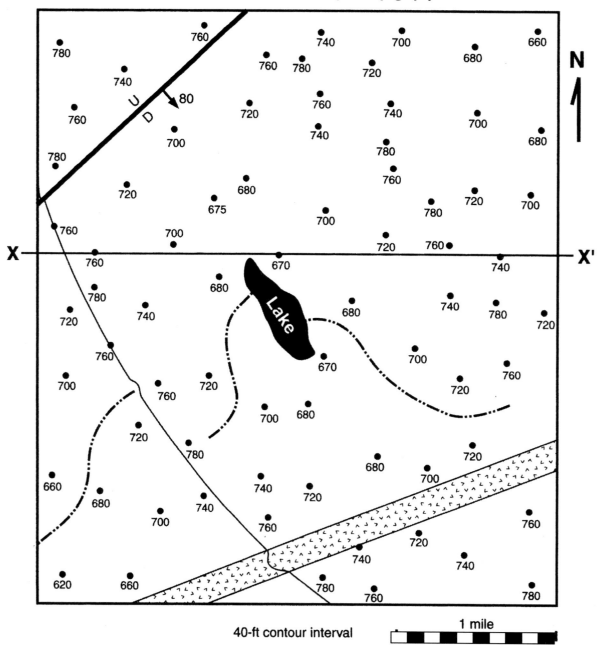

40-ft contour interval   1 mile

PART 3: Identify the hand specimens corresponding to units A, B, C, D and E. Your instructor will supply one set of samples for the entire class. Record the rock names on the answer sheet at the end of this chapter.

PART 3: Rock Identification

Rock unit A = _____

Rock unit B = _____

Rock unit C = _____

Rock unit D = _____

Rock unit E = _____

PART 4: Based on your rock identifications in PART 3 and using the appropriate rock patterns from Figure 10.1, draw the patterns for rock units A, B, D and E on the map in Figure 10.3.

FIGURE 10.3 Base Map for Geology

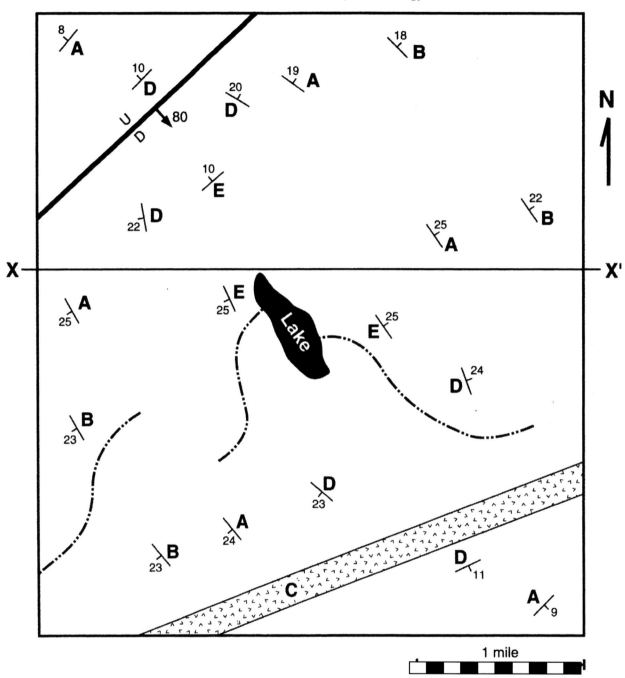

Figures from Physical Geology Exercises, 2nd edition by Jennifer Hinds. Copyright © 2004 by Kendall Hunt Publishing Company. Reprinted by permission.

Physical Geology Laboratory Exercises (EEES 1020)

PART 5: Draw a topographic profile along X-X' using the graph in Figure 10.4. Also in this figure, fill in the underlying geology, including both unit contacts and the appropriate rock pattern.

FIGURE 10.4 Graph for Topgraphic Profile and Geologic Cross Section Common Along X-X'

Figures from Physical Geology Exercises, 2nd edition by Jennifer Hinds. Copyright © 2004 by Kendall Hunt Publishing Company. Reprinted by permission.

PART 6: Answer the questions on the topography and geology of the map area.

PART 6: Questions

1.   What is the main topographic feature in Figure 10.2?

        _____

2.   What is the main geologic structure in Figure 10.3?

        _____

3.   What is the relationship between the topography and geologic structure in the map area?

        _____

        _____

        _____

4.   Which rock unit is least resistant to erosion? _____

        Explain why: _____

        _____

        _____

5.   Which rock unit is most resistant to erosion? _____

        Explain why:_____

        _____

6.   What is the topographic relief in the map area? _____

7.   What is the gradient (in feet per mile) of the stream in the southwest corner of the map?

        _____

8.   What is the oldest rock unit? _____

        What is the youngest rock unit? _____

9.   What kind of fault is shown in the northwest corner of the map?

_____

10.   Describe the geologic history of the map area using relative dating.

_____

_____

_____

_____

_____

_____

_____

_____

_____

_____

_____

_____

Chapter 11

Optional Exercise: Continental Glaciation

As a student at the University of Toledo, you are surrounded by evidence of **continental glaciation** from the last Ice Age. Between about 2 million and 10,000 years ago (i.e., during the Pleistocene Epoch of the Quaternary Period) large portions of the North American and Eurasian continents, including the Toledo area, were at times covered by vast sheets of ice up to 1000's of feet thick. These **continental glaciers** repeatedly advanced (grew by the addition of snow and ice during colder periods) and retreated (melted back during warmer periods).

A summary of the major landforms resulting from continental glaciation is shown in Figure 11.1. Deposits of gravel, sand and/or mud known as **moraines** are found at the lobate margins of ice sheets where they form terminal moraines (TM), recessional moraines (RM), and interlobate moraines (IM). Ground moraine (GM) is deposited beneath the moving glacier and may be locally shaped into drumlines (DR). Meltwater from the ice develops an outwash plain (OP) consisting of sand and gravel deposited by braided streams (BS) issuing from the glacier. Sinuous ridges called eskers (E) result from gravel and sand deposited on the floor of ice tunnels (T) cut by water flowing underneath a glacier. Ice blocks (IB), which have broken off from the glacier, become partly covered with outwash plain or ground moraine sediments, and develop kettle pits and lakes (K). Ice margin lakes (ML) form along the glacier front and receive sediments brought in by meltwater, which accumulate as deltas (D) or are dispersed over the lake bottom. These lake deposits form, respectively, delta kames (DK) of gravel and sand, and lacustrine plains (LP) of sand and mud.

FIGURE 11.1 Marginal Landforms of Continental Glaciers

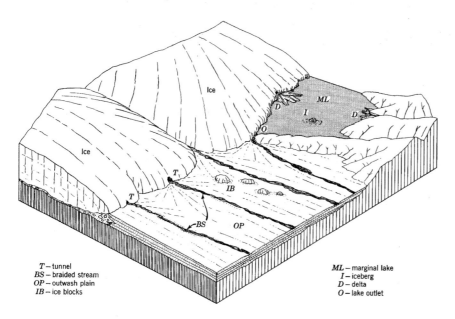

T — tunnel
BS — braided stream
OP — outwash plain
IB — ice blocks

ML — marginal lake
I — iceberg
D — delta
O — lake outlet

DURING GLACIAL ADVANCE

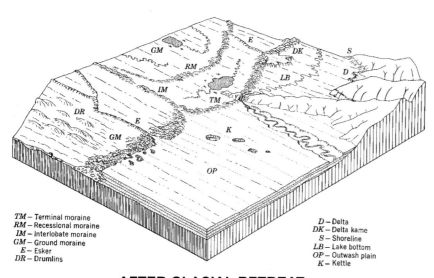

TM — Terminal moraine
RM — Recessional moraine
IM — Interlobate moraine
GM — Ground moraine
E — Esker
DR — Drumlins

D — Delta
DK — Delta kame
S — Shoreline
LB — Lake bottom
OP — Outwash plain
K — Kettle

AFTER GLACIAL RETREAT

From Introduction to Physical Geography by Arthur N. Strahler, 1965.

Exercise 11: Optional Exercise: Continental Glaciation

An especially good place to see some of the landforms produced by continental glaciation is in southern Michigan about 50 miles northwest of Toledo. Part of this area is shown in Topographic Map 3 (Jackson, Michigan). Using this map and the answer sheet provided at the end of this chapter, answer the following questions.

1. Study the area shown on the map, and outline and label each of the following glacial landforms: (a) esker, (b) terminal moraine, (c) ground moraine, and (d) kettle lakes.

2. Based on these landforms, in what direction did the glacier advance? Draw an arrow on the map showing this direction.

3. What do the many small areas enclosed by contours with hachure marks represent both in terms of surface elevations and glacial landforms?

   Question 3: _____

   _____

   _____

   _____

   _____

   _____

   _____

4. The poorly developed stream drainage and extensive wetlands (swamps and marshes) seen in the map are typical of areas that experienced continental glaciation. To emphasize the interconnected and poorly integrated character of this drainage, use a red pencil to trace the Grand River and its tributaries channels.

5. Based on both the drainage patterns and landforms, would you say the map area has experienced a lot or only a little erosion since the end of the Ice Age? Explain your reasoning.

continued

Exercise 11: Optional Exercise: Continental Glaciation, continued

Question 5: _____

_____

_____

_____

_____

_____

_____

Hand in the map and answer sheet to your instructor. These will be graded and returned to you.

Topographic Contour Maps

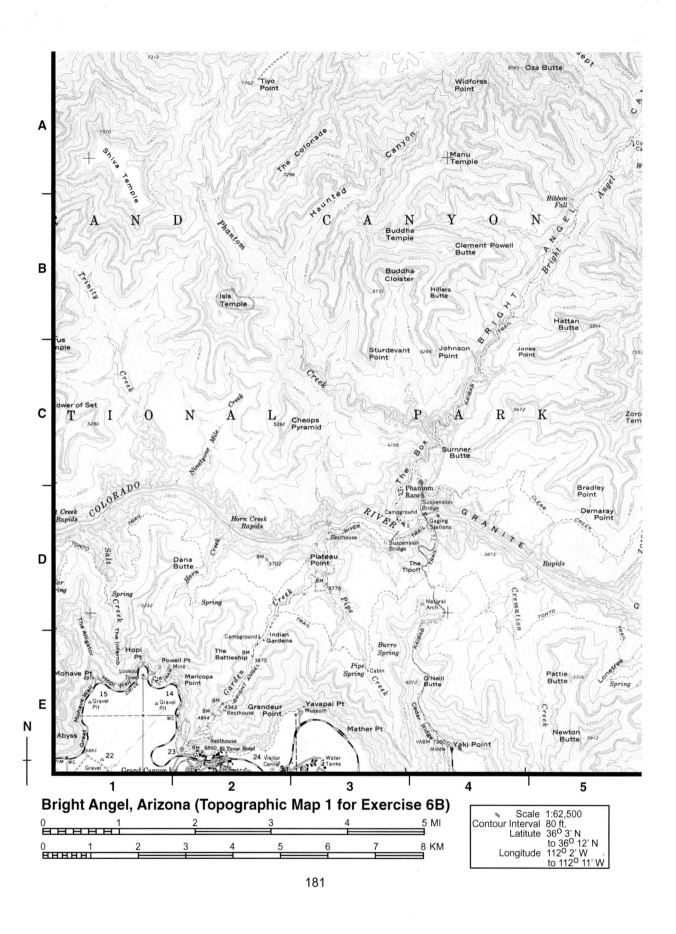

Bright Angel, Arizona (Topographic Map 1 for Exercise 6B)

0   1   2   3   4   5 MI

0   1   2   3   4   5   6   7   8 KM

Scale	1:62,500
Contour Interval	80 ft.
Latitute	36° 3' N
	to 36° 12' N
Longitude	112° 2' W
	to 112° 11' W

181

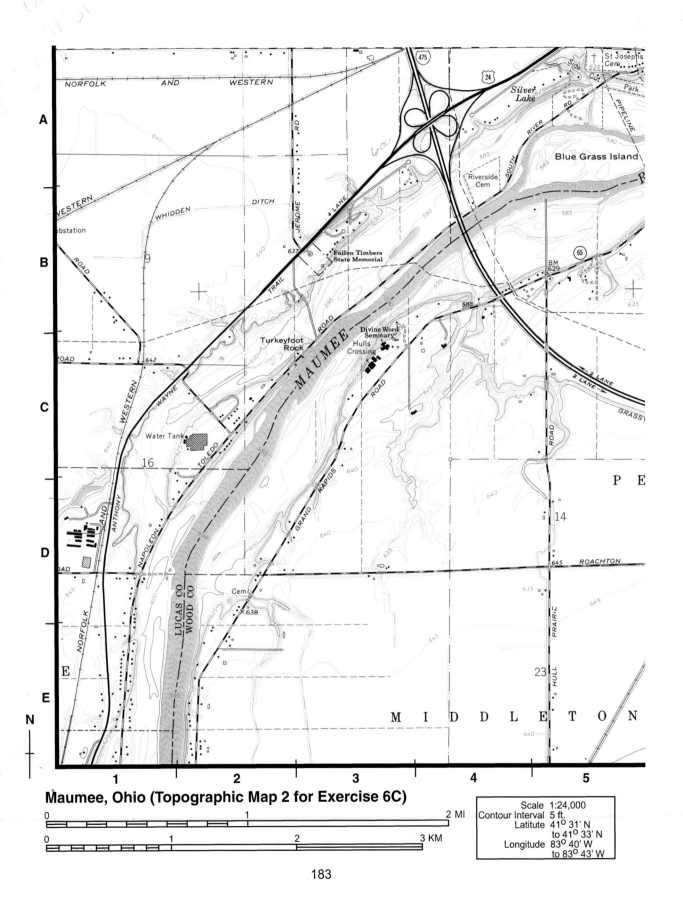

Maumee, Ohio (Topographic Map 2 for Exercise 6C)

0			1		2 MI

0		1		2	3 KM

Scale 1:24,000
Contour Interval 5 ft.
Latitude 41° 31' N
to 41° 33' N
Longitude 83° 40' W
to 83° 43' W

183

Topographic Contour Maps

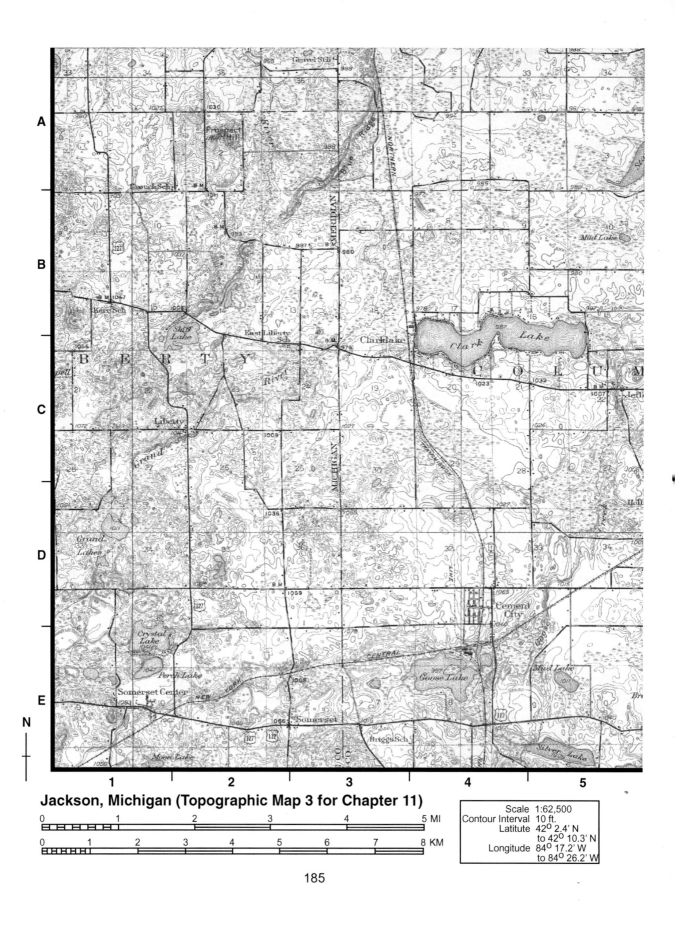

Jackson, Michigan (Topographic Map 3 for Chapter 11)

0	1	2	3	4	5 MI

0	1	2	3	4	5	6	7	8 KM

Scale 1:62,500
Contour Interval 10 ft.
Latitute 42° 2.4' N
to 42° 10.3' N
Longitude 84° 17.2' W
to 84° 26.2' W

185